Second
edition

Bridging GCSE and A-level Maths

Student Book

Mark Rowland

Contents

William Collins' dream of knowledge for all began with the publication of his first book in 1819.

A self-educated mill worker, he not only enriched millions of lives, but also founded a flourishing publishing house. Today, staying true to this spirit, Collins books are packed with inspiration, innovation and practical expertise. They place you at the centre of a world of possibility and give you exactly what you need to explore it.

Collins. Freedom to teach.

An imprint of HarperCollins*Publishers*
The News Building
1 London Bridge Street
London
SE1 9GF

Browse the complete Collins catalogue at
www.collins.co.uk

© HarperCollins*Publishers* Limited 2017

10 9 8 7 6 5 4 3 2 1

978-0-00-820501-0

Mark Rowland asserts his moral right to be identified as the author of this work.

British Library Cataloguing in Publication Data
A catalogue record for this publication is available from the British Library.

Commissioned by Jennifer Hall
Project editor Amanda Redstone
Project managed by Emily Hooton
Edited by Caroline Petherick
Proofread by Karen Williams
Answers checked by Steven Matchett
Cover design by We are Laura
Internal design by Graham Brasnett
Typesetting by Jouve India Private Limited
Illustrations by Kathy Baxendale and Ann Paganuzzi
Production by Rachel Weaver
Printed and bound by Grafica Veneta Spa, Italy

Introduction

Welcome to Collins Bridging GCSE and A Level Maths Student Book. This book helps you to progress smoothly onwards from GCSE Maths with detailed examples and plenty of practice in the key areas needed for success at A Level.

Revisiting GCSE

Consolidate your knowledge of difficult GCSE topics with graded worked examples.

Moving on to A Level

Find out how the strategies you learnt at GCSE are extended and explored with A Level worked examples.

Key points

Look out for the blue 'Key points' boxes that highlight the most important things to remember for each topic.

Handy hints and A level Alerts

Find valuable 'Handy hint' boxes throughout the book and understand methods that are specific to A Level with 'A Level Alert!' boxes.

Common errors and Checkpoints

Avoid the common misconceptions that students regularly make with 'Common errors' boxes and discover useful ways to check your workings with 'Checkpoint' boxes.

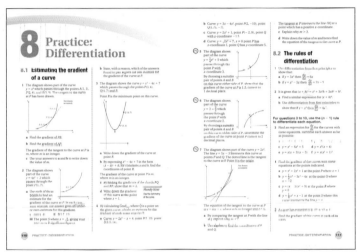

Practice section

Increase your confidence and improve your skills with comprehensive practice sections, packed with questions, dedicated to each topic. Look out for questions that have the problem solving icon (PS). These questions test your ability to reason and think your way through a problem.

Exam practice

Ensure you are ready to start your A Level course by taking the tear-out exam paper at the back of the book.

Answers

Find all the answers to the practice exercises at the back of the book. The exam paper answers are available on request from education@harpercollins.co.uk

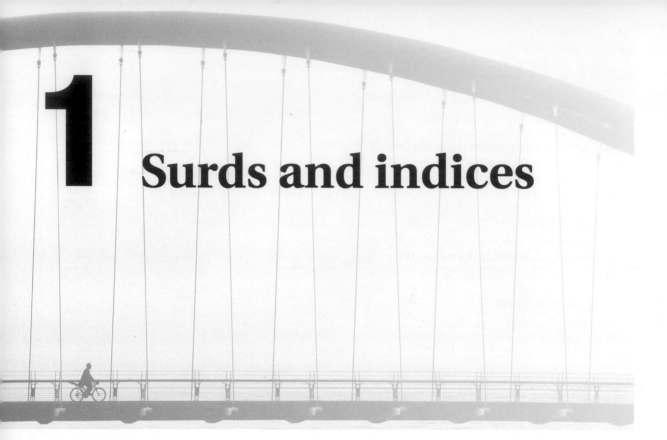

1 Surds and indices

1.1 Surds

What you should already know:

- how to simplify a surd such as $\sqrt{8}$.
- how to rationalise the denominator of a fraction such as $\dfrac{10}{\sqrt{5}}$.

In this section you will learn:

- how to use the rules of surds to simplify more complex expressions.
- how to rationalise more complex denominators in fractions.
- how to calculate with numbers and expressions involving surds.

> **Revisiting GCSE** >>

If a is a positive integer but $\sqrt{a}$ is not an integer then $\sqrt{a}$ is a **surd**. For example, $\sqrt{8}$ is a surd because a decimal approximation for $\sqrt{8}$ is $2.828\,427\,125$, which is not an integer. $\sqrt{9}$, on the other hand, is not a surd because its exact value is the integer 3.

You need to know the rules of surds:

Handy hint

You should be familiar with the square numbers: 1, 4, 9, 16, 25, 36…

Handy hint

You can assume $\sqrt{9}$ means the *positive* square root of 9.

> **Key point** >

The rules of surds are: $\sqrt{ab} = \sqrt{a} \times \sqrt{b}$, $\sqrt{\dfrac{a}{b}} = \dfrac{\sqrt{a}}{\sqrt{b}}$ (where $b \neq 0$).

These rules allow you to break down a surd. For example: $\sqrt{8} = \sqrt{4 \times 2} = \sqrt{4} \times \sqrt{2} = 2\sqrt{2}$

It is not possible to break down $\sqrt{2}$ any further using integers.
In **simplified surd form**, $\sqrt{8} = 2\sqrt{2}$.

GCSE Example 1

a Express $\sqrt{12} + \sqrt{48}$ in the form $k\sqrt{3}$ where k is an integer.

b Simplify $\dfrac{21}{\sqrt{3}}$ by rationalising the denominator.

Working

a You need to re-write $\sqrt{12}$ using factors of 12.

$$\sqrt{12} = \sqrt{4 \times 3}$$
$$= \sqrt{4} \times \sqrt{3}$$
$$= 2\sqrt{3}$$

Re-write $\sqrt{48}$ using factors of 48.

48 has lots of factors. Look for the *greatest* square number which is a factor of 48 (that is, 16).

$$\sqrt{48} = \sqrt{16 \times 3}$$
$$= \sqrt{16} \times \sqrt{3}$$
$$= 4\sqrt{3}$$

So $\sqrt{12} + \sqrt{48} = 2\sqrt{3} + 4\sqrt{3}$
$$= 6\sqrt{3} \text{ (so } k = 6\text{)}$$

b The denominator of $\dfrac{21}{\sqrt{3}}$ is the surd $\sqrt{3}$.

You need to find a fraction with the same value as $\dfrac{21}{\sqrt{3}}$ which does not have a surd in its denominator.

Since $\sqrt{3} \times \sqrt{3} = 3$, you multiply the numerator and denominator of this fraction by $\sqrt{3}$.

$$\frac{21}{\sqrt{3}} = \frac{21 \times \sqrt{3}}{\sqrt{3} \times \sqrt{3}}$$
$$= \frac{21\sqrt{3}}{3}$$
$$\frac{21}{3}\sqrt{3}$$
$$= 7\sqrt{3}$$

> **Key point**

$\sqrt{a} \times \sqrt{a} = a$ for any number $a > 0$.

> **Moving on to A Level**

At A Level, you will need to be able to simplify expressions involving surds.

A Level Example 2

Express $\dfrac{6 - \sqrt{6}}{3 + \sqrt{6}}$ in the form $a + b\sqrt{6}$ where a and b are integers to be stated.

Working

You need to rationalise the fraction $\dfrac{6 - \sqrt{6}}{3 + \sqrt{6}}$.

This is done by multiplying numerator and denominator by $(3 - \sqrt{6})$.

$$\frac{6 - \sqrt{6}}{3 + \sqrt{6}} = \frac{(6 - \sqrt{6}) \times (\mathbf{3} - \sqrt{\mathbf{6}})}{(3 + \sqrt{6}) \times (\mathbf{3} - \sqrt{\mathbf{6}})}$$

Simplify the numerator:
$$(6 - \sqrt{6}) \times (3 - \sqrt{6}) = 18 - 6\sqrt{6} - 3\sqrt{6} + \sqrt{6} \times \sqrt{6}$$

Combine like surds:
$$= 18 - 9\sqrt{6} + 6$$
$$= 24 - 9\sqrt{6}$$

Simplify the denominator:
$$(3 + \sqrt{6})(3 - \sqrt{6}) = 3 \times 3 - 3\sqrt{6} + 3\sqrt{6} - \sqrt{6} \times \sqrt{6}$$
$$= 9 - 6$$
$$= 3$$

So $\dfrac{6 - \sqrt{6}}{3 + \sqrt{6}} = \dfrac{24 - 9\sqrt{6}}{3}$

$$= \dfrac{24}{3} - \dfrac{9\sqrt{6}}{3}$$
$$= 8 - 3\sqrt{6}$$

$8 - 3\sqrt{6}$ looks like $a + b\sqrt{6}$ where $a = 8$, $b = -3$.

Key point

To rationalise the denominator of a fraction of the form $\dfrac{N}{a + \sqrt{b}}$, multiply top and bottom by $(a - \sqrt{b})$.

To rationalise the denominator of a fraction of the form $\dfrac{N}{a - \sqrt{b}}$, multiply top and bottom by $(a + \sqrt{b})$.

Taking it further

Surds appear in many areas of A Level Maths, such as solving quadratic equations and calculating distances between points.

1.2 Indices

What you should already know:

- how to simplify expressions such as $(a^2b)^2$.
- how to calculate values such as 3^{-2} or $4^{\frac{1}{2}}$.

In this section you will learn:

- how to work out the value of more complicated expressions using the rules of indices.

Revisiting GCSE

At GCSE, you would have used various rules of indices to simplify an expression. For example, in the expression $a^2 \times a^3$ you can add the indices together so that $a^2 \times a^3$ simplifies to a^5.

This works because $a^2 \times a^3 = (a \times a) \times (a \times a \times a)$
$$= a^5$$

Key point

Here are some rules of indices:

1 $a^m \times a^n = a^{m+n}$
2 $\dfrac{a^m}{a^n} = a^{m-n}$ for a not zero
3 $(a^m)^n = (a^n)^m = a^{mn}$
4 $a^n \times b^n = (ab)^n$
5 $\dfrac{a^n}{b^n} = \left(\dfrac{a}{b}\right)^n$ for b not zero

Handy hint

Rule **3** means $(a^m)^n$ and $(a^n)^m$ are equal to each other and are also equal to a^{mn}.

GCSE Example 3

Simplify $(2p^3q^2)^2$.

Working

You can write rule **4** as $(ab)^n = a^n \times b^n$.

So $\qquad\qquad (2p^3q^2)^2 = 2^2 \times (p^3)^2 \times (q^2)^2$

Use rule **3** on each bracket: $= 4 \times (p^{3 \times 2}) \times (q^{2 \times 2})$

Simplify the indices: $\qquad = 4p^6q^4$

So $(2p^3q^2)^2$ simplifies to $4p^6q^4$.

Handy hint

Rule **4** can be extended to three or more terms.

You will also have met the **definition** of negative and fractional indices.

For example, 2^{-3} means $\dfrac{1}{2^3}$ and so 2^{-3} has value $\dfrac{1}{8}$.

Similarly, $4^{\frac{1}{2}}$ means the positive square root of 4, and so $4^{\frac{1}{2}}$ has value 2.

$3^0 = 1$ and, in general, $a^0 = 1$ for any number a.

GCSE Example 4

Find the value of $3^{-2} \times 8^{\frac{1}{3}}$.

Working

You apply the definitions to each expression.

$3^{-2} = \dfrac{1}{3^2} \qquad\qquad 8^{\frac{1}{3}} = \sqrt[3]{8}$

$\qquad = \dfrac{1}{9} \qquad\qquad\qquad = 2$

So $3^{-2} \times 8^{\frac{1}{3}} = \dfrac{1}{9} \times 2$

$\qquad\qquad\qquad = \dfrac{2}{9}$

Moving on to A Level

Key point

Here are some general definitions which you need to know for A Level Maths.

6 $a^{-n} = \dfrac{1}{a^n}$ where n is any integer.

7 $a^{\frac{1}{n}} = \sqrt[n]{a}$ where n is any positive integer.

At A Level, you will be expected to manipulate expressions involving indices.

A Level Example 5

a Find the value of t for which $8^t = \frac{1}{16}$.

b Express $\dfrac{6\sqrt{y}+1}{2y^2}$ in the form $3y^p + qy^r$, for p, q and r constants to be stated.

Working

a You need to express 8 and 16 as powers of 2.

$$8 = 2^3 \text{ and } 16 = 2^4$$

So the equation $8^t = \frac{1}{16}$ can be written as $\left(2^3\right)^t = \frac{1}{2^4}$.

Use rules **3** and **6**: $\qquad\qquad\qquad\qquad\qquad 2^{3t} = 2^{-4}$

Base numbers are equal so equate indices: $\; 3t = -4$

So $\qquad\qquad\qquad\qquad\qquad\qquad\qquad\qquad t = -\dfrac{4}{3}$

> **Checkpoint**
>
> If your calculator has a 'power' key $\boxed{x^{\square}}$, use it to check that $8^{-\frac{4}{3}} = \frac{1}{16}$.

b You need to split the fraction up into two terms.

$$\frac{6\sqrt{y}+1}{2y^2} = \frac{6\sqrt{y}}{2y^2} + \frac{1}{2y^2}$$

Use rule **7**: $\qquad\qquad\qquad\qquad = \dfrac{6y^{\frac{1}{2}}}{2y^2} + \dfrac{1}{2y^2}$

> **Common error**
>
> $\frac{1}{2y^2}$ does **not** simplify to $2y^{-2}$.

Separate the number fractions from the algebraic fractions:
$$= \frac{6}{2} \times \frac{y^{\frac{1}{2}}}{y^2} + \frac{1}{2} \times \frac{1}{y^2}$$

Use rules **2** and **6**: $\qquad\qquad = 3 \times y^{\frac{1}{2}-2} + \frac{1}{2} \times y^{-2}$

Simplify: $\qquad\qquad\qquad\qquad = 3y^{-\frac{3}{2}} + \frac{1}{2}y^{-2}$

> **Handy hint**
>
> You must list the values of p, q and r, as directed by the question.

$3y^{-\frac{3}{2}} + \frac{1}{2}y^{-2}$ equals $3y^p + qy^r$ provided $p = -\dfrac{3}{2}$, $q = \dfrac{1}{2}$ and $r = -2$.

Key point

$$a^{\frac{m}{n}} = \left(\sqrt[n]{a}\right)^m$$
$$= \sqrt[n]{a^m}$$

When calculating $a^{\frac{m}{n}}$ it is usually easier to use the result $\left(\sqrt[n]{a}\right)^m$ rather than $\sqrt[n]{a^m}$.

Taking it further

Indices appear in many A Level Maths topics, but in particular you will need to be confident in using them when studying differentiation and integration.

2 Algebra 1

2.1 Basic algebra

What you should already know:

- how to simplify expressions such as $a(a + 4) - (a + 1)^2$ or $(2a^3b)^2$.
- how to rearrange a formula such as $A = \pi r^2$ to make r the subject.

In this section you will learn:

- how to deal with more complicated expressions and formulae.

》》 **Revisiting GCSE** 》》

At GCSE, you practised expanding brackets, factorising and simplifying expressions,

GCSE Example 1

a Factorise fully $12a^5b + 6a^2b^3$.

b Rearrange $A = \dfrac{h(a + b)}{2}$ to make b the subject of the formula.

Working

a You look for common factors of numbers and terms.

In the expression $12a^5b + 6a^2b^3$

the numbers are 12 and 6:   their highest *common* factor is 6 because $12 = 6 \times 2$.

the powers of a are a^5 and a^2:   their highest *common* power is a^2 because $a^5 = a^2 \times a^3$.

the powers of b are b and b^3:   their highest *common* power is b^1 because $b^3 = b^1 \times b^2$.

So the highest common factor of the expression is $6a^2b$.

In fully factorised form, $12a^5b + 6a^2b^3 = 6a^2b(2a^3 + b^2)$.

> **Handy hint**
>
> The smaller of the two powers gives the common factor.

> **Checkpoint**
>
> Multiplying out the brackets gives you the expression you started with.

b Formula to rearrange:

$$A = \frac{h(a+b)}{2}$$

Multiply both sides by 2:

$$2A = h(a+b)$$

Divide both sides by h:

$$\frac{2A}{h} = a + b$$

Subtract a from both sides:

$$\frac{2A}{h} - a = b$$

So $b = \frac{2A}{h} - a$.

▶ Moving on to A Level ▶▶▶▶

At A Level, as well as being confident in expanding and factorising expressions, you will also need to be able to express an equation in a particular form for use in another part of a question. There is usually more than one way of doing this.

A Level Example 2

Express $\frac{8x^5 + 6x^2}{2x^2}$ in the form $Ax^3 + B$, stating the value of the constants A and B.

Working

In one method you fully factorise the numerator.

$$8x^5 + 6x^2 = 2x^2(4x^3 + 3)$$

So

$$\frac{8x^5 + 6x^2}{2x^2} = \frac{2x^2(4x^3 + 3)}{2x^2}$$

$$= 4x^3 + 3$$

Compare $4x^3 + 3$ with $Ax^3 + B$: $A = 4$, $B = 3$.

Alternatively, you can divide each term in the numerator by $2x^2$.

$$\frac{8x^5 + 6x^2}{2x^2} = \frac{8x^5}{2x^2} + \frac{6x^2}{2x^2}$$

$$= 4x^3 + 3$$

You may also be required to find a formula and then rearrange it, where the required letter appears on *both* sides of the formula.

A Level Example 3

The diagram shows a right-angled triangle ABC, where $BC = p$ and $AC = p + 1$.

If $s = \sin \hat{A}$ find an expression for p in terms of s.

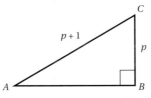

Working

You need to use right-angled trigonometry.

$$\sin \hat{A} = \frac{\text{opposite}}{\text{hypotenuse}} \text{ so } s = \frac{p}{p+1}$$

Formula to rearrange: $s = \frac{p}{p+1}$

Multiply both sides by $(p+1)$: $s(p+1) = p$

Expand brackets: $sp + s = p$

Gather terms in p to one side and terms in only s to the other:

$$sp - p = -s$$

Factorise the left-hand side: $(s-1)p = -s$

Divide both sides by $(s - 1)$: $p = \dfrac{-s}{s - 1}$

So an expression for p in terms of s is $p = \dfrac{s}{1 - s}$.

⟫ Taking it further ⟫⟫⟫

Being able to manipulate algebraic expressions quickly is an important skill in A Level Maths. Most questions depend on these techniques!

2.2 Solving linear equations

What you should already know:

- how to solve an equation such as $2(3a + 1) = a + 7$.
- how to solve a pair of simultaneous equations such as $2x + 3y = 11$ and $3x - 2y = 10$.

In this section you will learn:

- how to solve a complex problem by breaking it down into steps.

⟫ Revisiting GCSE ⟫⟫

At GCSE, you learnt how to solve an equation by applying reverse operations to each side of the equation.

GCSE Example 4

Solve these equations.

a $\dfrac{2x}{3} + 1 = 7$

b $\dfrac{3t}{t + 3} = 2$

Working

a You need to 'undo' the equation by applying reverse operations to each side, so that adding becomes subtracting, dividing becomes multiplying, and so on.

Equation to solve: $\dfrac{2x}{3} + 1 = 7$

Subtract 1 from both sides: $\dfrac{2x}{3} = 6$

Multiply both sides by 3: $2x = 18$

Divide both sides by 2: $x = 9$

The solution is $x = 9$.

b You need to multiply both sides by the denominator $(t + 3)$.

Equation to solve: $\dfrac{3t}{(t + 3)} = 2$

Multiply both sides by $(t + 3)$: $3t = 2(t + 3)$

Expand the bracket: $3t = 2t + 6$

Subtract $2t$ from both sides: $t = 6$

So the solution is $t = 6$.

Here is an example of an A Level question where you have to think your way through the problem.

A Level Example 5

The diagram shows a rectangle $ABCD$.
$AB = x$ cm and $BC = (2x + 2)$ cm.
The perimeter of this rectangle is 34 cm.
Find the length of the diagonal AC.

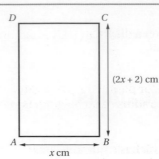

Working

Angle $ABC = 90°$ so you can use Pythagoras' theorem to find the length AC.

> **A Level Alert!**
> You need to develop this kind of thinking skill when solving a problem.

To use Pythagoras, you will first need to find the lengths of the sides AB and BC.

To find the value of x, set up and solve an equation using the given information.

$ABCD$ is a rectangle, so its perimeter in terms of x is:

$$x + (2x + 2) + x + (2x + 2)$$
$$= 6x + 4 \text{ cm}$$

The numerical value of the perimeter is 34 cm so $6x + 4 = 34$.

Subtract 4 from both sides: $6x = 30$

Divide both sides by 6: $x = 5$

So $AB = 5$ cm and $BC = 2(5) + 2$
$$= 12 \text{ cm}$$

You can now use Pythagoras' theorem to find the length of the diagonal AC.

$$AC^2 = AB^2 + BC^2$$
$$= 5^2 + 12^2$$
$$= 25 + 144$$
$$= 169$$

So $AC = \sqrt{169}$
$$= 13$$

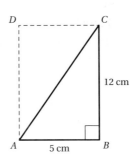

> **A Level Alert!**
> Units are important. Make sure you include them in your final answer where appropriate.

The diagonal AC has length 13 cm.

Taking it further 〉〉〉〉〉

Equation-solving features in almost every aspect of A Level Maths. You may need to be able to form your own equations and solve them without being told exactly what to do at each step.

2.3 Forming expressions

What you should already know:

* how to form an expression from a diagram.

In this section you will learn:

* how to work with two variables on a diagram.

 Revisiting GCSE

At GCSE, you worked with formulae for perimeters, areas and volumes of various shapes. For example, the volume V of a cylinder with base radius r and height h is given by the formula $V = \pi r^2 h$.

GCSE Example 6

The diagram shows a field $ABCDE$ which is a quarter-circle ABE, centre B, radius $2x$ metres, joined to a square $BCDE$.

Find an expression for the perimeter of the field.

Give your answer in the form $(a + \pi)x$ where a is an integer.

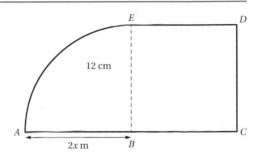

Working

You need to find an expression for the length of each edge of the field.

The radius of the quarter-circle is $2x$ so the arc length

$AE = \dfrac{1}{4} \times 2\pi(2x)$

$= \pi x.$

$BE = 2x$ because BE is a radius of the quarter-circle.

Since $BCDE$ is a square, $BC = CD = DE = 2x.$

The perimeter of the field $= 2x + 2x + 2x + 2x + \pi x$

$\qquad\qquad\qquad\qquad\quad = 8x + \pi x$

Factorise: $\qquad\qquad\qquad = (8 + \pi)x$ metres

> **Handy hint**
>
> The circumference of a circle radius r is $2\pi r$.

> **Common error**
>
> $8x + \pi x$ is **neither** $8\pi x$ **nor** $8\pi x^2$

You can turn this expression into an algebraic equation by writing $P = (8 + \pi)x$ where P is the perimeter of the field (in metres).

 Moving on to A Level

At A Level, you may have to rewrite an expression which uses two variables as one which uses a single variable.

A Level Example 7

The solid shape shown in the diagram has a uniform cross-section.

Triangle ABC is right-angled where $AB = 4x$ and $BC = 3x$.

$AE = y$ and all lengths are in cm.

The volume V cm^3 of this shape is 60 cm^3.

a Find y in terms of x.

b Hence show that the area of the rectangular face $ACDE$ is $\dfrac{50}{x}$ cm^2.

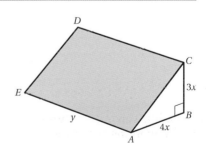

Working

a The solid shape has a constant cross-section, so, working in cm³:

$V = (\text{Area of triangle } ABC) \times y$

$\quad = \frac{1}{2}(4x)(3x)y$

$\quad = 6x^2y$

The numerical value of the volume is 60 cm³, so: $6x^2y = 60$

Divide each side by $6x^2$: $\qquad\qquad\qquad\qquad y = \dfrac{60}{6x^2}$

So $\qquad\qquad\qquad\qquad\qquad\qquad\qquad\qquad y = \dfrac{10}{x^2}$

Handy hint

You must use all the given information, in this case $V = 60$ cm³.

b You can use Pythagoras' theorem to find an expression for the length of AC.

By Pythagoras: $AC^2 = AB^2 + BC^2$

$\qquad\qquad\qquad\quad = (4x)^2 + (3x)^2$

$\qquad\qquad\qquad\quad = 16x^2 + 9x^2$

$\qquad\qquad\qquad\quad = 25x^2$

So $\qquad\qquad AC = \sqrt{25x^2}$

$\qquad\qquad\qquad\quad = 5x$

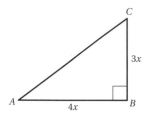

Area of rectangle $ACDE = 5x \times y$

$\qquad\qquad\qquad\qquad\quad = 5x \times \dfrac{10}{x^2}$

$\qquad\qquad\qquad\qquad\quad = \dfrac{50x}{x^2}$

$\qquad\qquad\qquad\qquad\quad = \dfrac{50}{x}$ cm², as required.

Handy hint

Part **b** is a 'Hence' question – you must use the answer $y = \dfrac{10}{x^2}$ found in part **a**.

Common error

$5x \times \dfrac{10}{x^2}$ does **not** simplify to $\dfrac{50x}{5x^3}$

Write this product as $\dfrac{5x}{1} \times \dfrac{10}{x^2}$ to avoid this error.

❯❯ Taking it further ❯❯❯❯❯

Being able to form expressions appears in practical problems when trying to maximise or minimise quantities such as surface areas or volumes.

3 Coordinate geometry 1

3.1 Straight-line graphs

What you should already know:

- $y = mx + c$ is the equation of a straight line with gradient m and y-intercept c.
- how to identify the gradient and y-intercept of a line using its equation and/or graph.
- how to tell if two lines are parallel or are perpendicular to each other.

In this section you will learn:

- how to sketch the graph of a straight line with equation $y = mx + c$ or $ay + bx = c$.
- how to interpret the gradient and y-intercept of a linear model.

>>> **Revisiting GCSE** >>>

At GCSE, you learnt that every straight-line graph has equation $y = mx + c$ where m (the coefficient of x) is the gradient and c (the constant term) is the y-intercept of the line.

This diagram shows a drawing of the line with equation $y = 2x + 3$.

For this line, the gradient is 2, which means starting from any point on the line, going 1 across and 2 up takes you to another point on the line.

The y-intercept is 3, which is where the line crosses the y-axis.

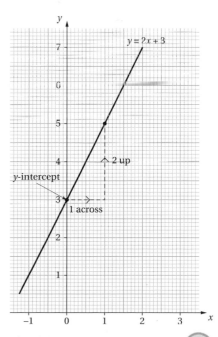

At GCSE, you learnt that two lines are:

- parallel if they have equal gradients
- perpendicular if the product of their gradients is -1.

The diagram shows two parallel lines, P and Q, with gradient m, and a third line R which is perpendicular to both P and Q.

Here is a GCSE question about parallel and perpendicular lines.

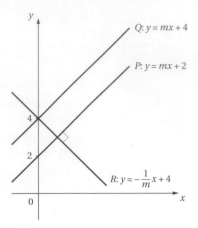

GCSE Example 1

These are the equations of three straight lines.

Line A: $y = \frac{4}{3}x + 3$

Line B: $y = \frac{3}{4}x - 2$

Line C: $4y + 3x = 12$

Two of these lines are perpendicular to each other.

a Find these two lines.

b Find an equation for the line L which is parallel to line C and which intersects line B on the y-axis.

Working

a You need to find the gradient of each line by comparing its equation with $y = mx + c$.

Line A: $y = \frac{4}{3}x + 3$ has gradient $\frac{4}{3}$

Line B: $y = \frac{3}{4}x - 2$ has gradient $\frac{3}{4}$

$\frac{4}{3} \times \frac{3}{4} = 1$ so line A and line B are not perpendicular.

Line C: Rearrange $4y + 3x = 12$ to make y the subject.

$$4y + 3x = 12$$

Subtract $3x$ from each side: $4y = 12 - 3x$

Divide all terms by 4: $\qquad y = 3 - \frac{3}{4}x$

Line C has gradient $-\frac{3}{4}$.

$\frac{4}{3} \times -\frac{3}{4} = -1$ so line A and line C are perpendicular.

b You need to find the gradient and the y-intercept of line L.

Line L and line C are parallel so the gradient of line L is $-\frac{3}{4}$.

Line L and line B have the same y-intercept, so the y-intercept of line L is -2.

An equation for line L is $y = -\frac{3}{4}x - 2$.

Handy hint

You can only read off the gradient and y-intercept of a line when its equation is in the form $y = mx + c$.

Common error

The gradient of the line $y = 3 - \frac{3}{4}x$ is **not** 3. In this form, the gradient is the coefficient of x.

Handy hint

A sketch is useful for checking that answers look reasonable.

You need to be able to **sketch** a line.

A sketch:

*   does not use scales on the axes.
*   should show only the main features of the graph such as its axis-crossing points.

For example, you can sketch the line L with equation $3y - 4x = 12$ by finding where L intersects each coordinate axis.

Let $x = 0$ to find the y-intercept: $3y - 4(0) = 12$

$$so \quad y = \frac{12}{3} = 4$$

Let $y = 0$ to find the x-intercept: $3(0) - 4x = 12$

$$so \quad x = \frac{12}{-4} = -3$$

Line L passes through the points $(0, 4)$ and $(-3, 0)$.

The diagram shows a sketch of line L.

You also have to interpret a line equation which describes a real-life problem. Such a line is called a **linear model**.

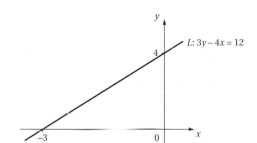

A Level Example 2

The depth of water h cm in a bath t minutes after the plug has been removed was modelled by the straight line equation $h = 19 - 9.5t$.

a   Sketch the graph of h against t for $t \geq 0$.

b   For this line describe, in context:

   i   what each axis crossing point represents.

   ii   what the gradient represents.

c   Explain why this model is not valid for $t > 2$.

Working

a   A graph of h against t means h is plotted on the vertical axis and t on the horizontal axis. The y and x axes in Figure 1 can be re-labelled as the h and t axes, as shown in Figure 2.

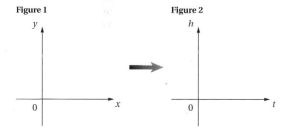

Figure 1

Figure 2

The line with equation $h = 19 - 9.5t$ has h-intercept 19 and gradient -9.5 (which means the line slopes downwards as t increases).

To complete the sketch you need to find where the line crosses the t-axis, which happens when $h = 0$.

Equation to solve: $0 = 19 - 9.5t$

Add $9.5t$ to each side: $19.5t = 19$

Divide each side by 9.5: $t = \dfrac{19}{9.5}$

$$= 2$$

The diagram shows the sketch of the graph of h against t.

b   **i**   The h-intercept 19 means the depth of the water when the plug was pulled was 19 cm.

       The t-intercept 2 means it took 2 minutes for the bath to empty.

   ii   The **negative** gradient -9.5 means the depth was **decreasing** at a rate of 9.5 cm per minute.

c   From the graph, when $t > 2$, h is negative, which has no physical meaning.

    The model is not valid after 2 minutes.

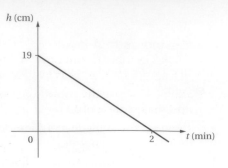

>> **Taking it further** >>>>>

You will deal with straight lines extensively in the Coordinate Geometry sections of A Level Maths. Straight lines are also fundamental to understanding the gradient of a curve in the study of differentiation.

3.2   Finding the equation of a line

What you should already know:

* how to calculate the gradient of a line using two points.
* how to find the equation of a line in the form $y = mx + c$ which passes through two points.

In this section you will learn:

* how to find the equation of any line using the formula $y - y_1 = m(x - x_1)$.
* how to use coordinates of points to solve geometrical problems.

>> **Revisiting GCSE** >>>

At GCSE, you learnt how to find the equation of a line by first calculating its gradient. The y-intercept was either given to you, or you could find it by using the coordinates of a point on the line.

Here is a GCSE-style question.

GCSE Example 3

Find the equation of the line L which passes through the points $A(2, 1)$ and $B(6, 9)$.

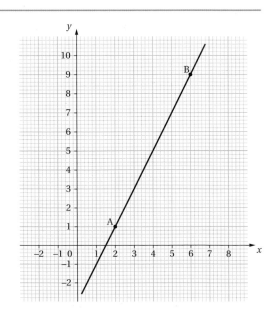

Working

You know that an equation for L is $y = mx + c$ where m is its gradient and c is its y-intercept.

You need to use the coordinates of points A and B to find the gradient of L.

Draw a right-angled triangle under the line AB.

The gradient $m = \dfrac{\text{height}}{\text{base}} = \dfrac{9-1}{6-2} = \dfrac{8}{4} = 2$.

So an equation for L is $y = 2x + c$.

You can use the coordinates of point A to find the value of c.

Since $A(2, 1)$ is on L, you know that $x = 2$, $y = 1$ must satisfy the equation $y = 2x + c$.

Line equation: $y = 2x + c$

Substitute: $x = 2, y = 1$, $1 = 2(2) + c$

So $c + 4 = 1$

Subtract 4 from each side: $c = 1 - 4$

$\qquad\qquad\qquad\qquad\quad = -3$

An equation for L is $y = 2x - 3$.

You should check your answer by using the coordinates of $B(6, 9)$.

When $x = 6$, $y = 2x - 3$

$\qquad\qquad = 2(6) - 3$

$\qquad\qquad = 9$ (checked)

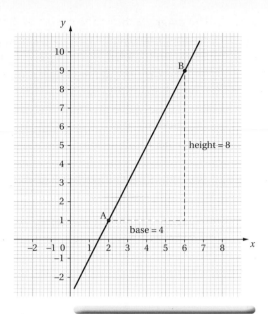

Handy hint

You could have used $B(6, 9)$ instead to find c.

Checkpoint

You can see from the diagram that:

- the gradient is positive

- the y-intercept is negative.

The answers $m = 2$, $c = -3$ make sense.

Key point

A line which passes through the points (x_1, y_1) and (x_2, y_2) has gradient $m = \dfrac{y_2 - y_1}{x_2 - x_1}$.

At A level, you may need to express an equation of a line in the form $ay + bx = c$, where a, b and c are **integers** (whole numbers) and to solve geometrical problems.

Handy hint

Gradient = 'the change in y divided by the change in x'.

If a line with equation $y = mx + c$ passes through the point (x_1, y_1), then $y_1 = mx_1 + c$ and so $c = y_1 - mx_1$.

This means the equation of the line can be written as $y = mx + (y_1 - mx_1)$ which can be re-expressed as $y - y_1 = m(x - x_1)$.

Key point

An equation for the line which passes through the point (x_1, y_1) and which has gradient m is $y - y_1 = m(x - x_1)$.

Handy hint

You will find this result helpful when working with lines at A Level.

Here is an A-level style question.

A Level Example 4

The line, L, passes through the points $P(1, 2)$ and $Q(-5, 7)$.

a Find an equation for L, giving your answer in the form $ay + bx = c$, where a, b and c are integers.

b Find the area of triangle OQR where O is the origin and R is the point where this line intersects the x-axis.

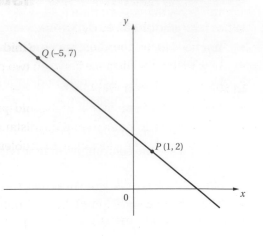

Working

a Sketch the line by making a rough plot of the points.

$P(1, 2)$ and $Q(-5, 7)$.

Gradient $m = \dfrac{y_2 - y_1}{x_2 - x_1}$ where $(x_1, y_1) = (1, 2)$ and

$(x_2, y_2) = (-5, 7)$.

So $m = \dfrac{7 - 2}{-5 - 1}$

$\quad = -\dfrac{5}{6}$

Use the formula $y - y_1 = m(x - x_1)$ with $(x_1, y_1) = (1, 2)$ to find an equation for L.

$y - 2 = -\dfrac{5}{6}(x - 1)$

Multiply each side by 6 to clear the fraction:

$$6y - 12 = -5(x - 1)$$

Open the bracket: $\quad 6y - 12 = -5x + 5$

Rearrange answer into required form: $\quad 6y + 5x = 17$

An equation for L is $6y + 5x = 17$.

b You need to find the x-intercept of L.

Let $y = 0$ to find the x-intercept: $\quad 6(0) + 5x = 17$

Solve for x: $\quad x = \dfrac{17}{5}(= 3.4)$

The area of triangle $OQR = \dfrac{1}{2} \times$ base $\times$ height

$\qquad = \dfrac{1}{2} \times 3.4 \times 7$

$\qquad = 11.9$ square units

> *Handy hint*
>
> You would get the same answer for m if instead you made $(x_1, y_1) = (-5, 7)$ and $(x_2, y_2) = (1, 2)$.

> *Handy hint*
>
> You need to clear the fraction because the equation must involve integers.

> *Handy hint*
>
> You would get the same equation if instead you used point $(x_2, y_2) = (-5, 7)$.
> Try this for yourself.

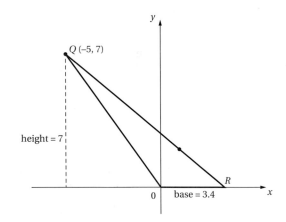

Taking it further

You need to be able to find and use equations of lines to solve complex problems in Coordinate Geometry.

3.3 Mid-points and distances

What you should already know:

- how to find the coordinates of the mid-point of the line joining two points by using a graph.
- how to find the distance between two points by using a graph.

In this section you will learn:

- how to use a formula to find the mid-point of a line joining points A and B.
- how to use a formula to find the distance between two points by using a graph.
- how to apply these formulae to problems.

▶▶ Revisiting GCSE ▶▶

At GCSE, you may have found the mid-point of a line joining two points, and the distance between them, by drawing a triangle under the line.

GCSE Example 5

The grid shows a line passing through the points $A(2, 1)$ and $B(8, 5)$.

a Find the distance AB, giving the answer in simplified surd form.

b Find the coordinates of the midpoint, M, of AB.

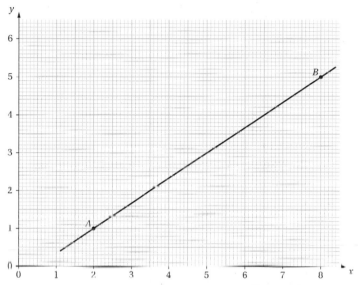

Working

a Draw a right-angled triangle underneath the line AB with base 6 and height 4.

By Pythagoras: $AB^2 = 6^2 + 4^2$
$$= 36 + 16$$
$$= 52$$

So $AB = \sqrt{52}$
$$= 2\sqrt{13}$$

b To go from A to B, you go 6 across and 4 up.

So, because M is the midpoint of AB, to go from A to M, you only need to go 3 across and 2 up.

The coordinates of M are $(2 + 3, 1 + 2) = (5, 3)$.

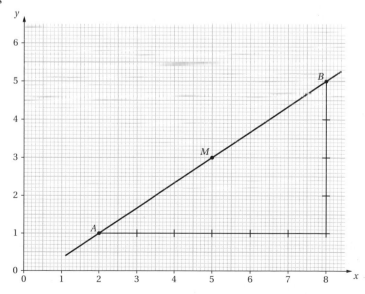

In A level Maths, it is more convenient to use formulae for finding distances and the coordinates of midpoints.

>> **Key point** >>>

For the points $A(x_1, y_1)$ and $B(x_2, y_2)$:

– the length of the line $AB = \sqrt{\left(x_2 - x_1\right)^2 + \left(y_2 - y_1\right)^2}$

– the midpoint M of the line AB has coordinates $\left(\dfrac{x_1 + x_2}{2}, \dfrac{y_1 + y_2}{2}\right)$.

You can use Pythagoras' theorem and similar triangles to prove these formulae.

In the diagram, M is the midpoint of the line AB.

The horizontal distance $AD = (x_2 - x_1)$ and

the vertical distance $DB = (y_2 - y_1)$.

By Pythagoras, $AB^2 = AD^2 + DB^2$.

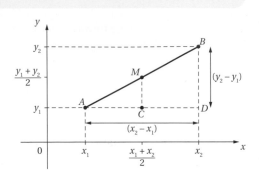

So $AB = \sqrt{\left(x_2 - x_1\right)^2 + \left(y_2 - y_1\right)^2}$.

Also, because $AB = 2 \times AM$ and corresponding angles in the right-angled triangles ACM and ADB are equal, $\triangle ADB$ is a scale factor 2 enlargement of $\triangle ACM$.

Hence $AD = 2 \times AC$ and so C is the midpoint of the horizontal line AD

This means the x-coordinate of C, and therefore of M, is $\dfrac{x_1 + x_2}{2}$.

Similarly, the y-coordinate of M is $\dfrac{y_1 + y_2}{2}$.

> *A Level Alert!*
>
> Circles can be centred anywhere, not just at the origin.

A Level Example 6

The diagram shows a circle, centre C which passes through the points $P(1, 3), Q(7, -1), R(11, 5)$ and S.

> *Handy hint*
>
> A diameter is any chord of a circle which passes through its centre.

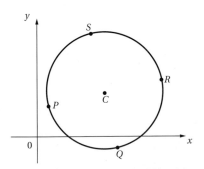

The lines PR and QS are diameters of this circle.

a   Find the area of this circle.

b   Find the coordinates of point S.

Working

a The area of the circle is πr^2 where r is its radius.

PR is a diameter of this circle, so $r = \frac{1}{2} \times PR$.

Use the distance formula with $(x_1, y_1) = (1, 3)$ and $(x_2, y_2) = (11, 5)$ to find PR.

$$PR = \sqrt{(x_2 - x_1)^2 + (y_2 - y_1)^2}$$
$$= \sqrt{(11 - 1)^2 + (5 - 3)^2}$$
$$= \sqrt{(10)^2 + (2)^2}$$
$$= \sqrt{104}$$

So the radius $r = \frac{1}{2} \times \sqrt{104}$
$$= \sqrt{26}$$

> **Common error**
>
> $\sqrt{10^2 + 2^2}$ is **not** equal to $10 + 2$.

The area of the circle $= \pi \times \left(\sqrt{26}\right)^2$
$$= 26\pi \text{ square units}$$

b Let the coordinates of S be (p, q).

To find these coordinates, you need to first find the coordinates of point C.

C is the midpoint of PR, because PR is a diameter of this circle, and C is its centre.

$P(1, 3)$, $R(11, 5)$ so the midpoint of PR has coordinates

$$\left(\frac{1 + 11}{2}, \frac{3 + 5}{2}\right) = (6, 4).$$

Hence, C has coordinates $(6, 4)$

Again, because QS is a diameter of this circle, C is also the midpoint of QS.

$Q(7, -1)$ and $S(p, q)$ so the midpoint of QS has coordinates

$$\left(\frac{7 + p}{2}, \frac{-1 + q}{2}\right).$$

Hence: $\left(\frac{7 + p}{2}, \frac{-1 + q}{2}\right) = (6, 4).$

Compare the x-coordinates: $\frac{7 + p}{2} = 6$

Multiply each side by 2: $7 + p = 12$

Subtract 7 from each side: $p = 5$

Similarly, by comparing y-coordinates, $q = 9$.

So the coordinates of S are $(5, 9)$.

> **A Level Alert!**
>
> You need to define symbols to represent the unknown quantities.

> **Handy hint**
>
> The midpoint formula can still be applied even though p and q are unknown.

> **Checkpoint**
>
> Use the distance formula to verify $QS = \sqrt{104}$.

Taking it further

You will use the formulae for mid-points and distances when solving Coordinate Geometry problems. The distance formula is used when finding the equation of a circle and when working with vectors.

3.4 Intersections of lines

What you should already know:

- how to use algebra to find the coordinates of the point where two lines intersect.

In this section you will learn:

- how to solve geometrical problems involving intersecting lines.

At GCSE, you learnt how to find the exact point where two lines intersect by solving a pair of simultaneous equations.

GCSE Example 7

The diagram shows the graph of $y = \frac{5}{2}x - 3$ and the graph of $4y + 3x = 10$.

These lines intersect at point P.

Find the exact coordinates of point P.

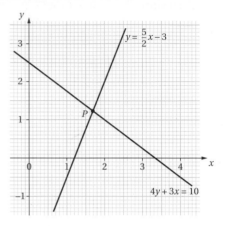

Working

It is not possible to read off exactly from the graph where these two lines meet.

You need to use algebra to find the exact coordinates of point P.

Solve the simultaneous equations $y = \frac{5}{2}x - 3$ ①

$$4y + 3x = 10 \qquad ②$$

by replacing the y term in equation ② with the expression $\left(\frac{5}{2}x - 3\right)$ from equation ①.

Equation ② becomes: $4\left(\frac{5}{2}x - 3\right) + 3x = 10$

Expand the bracket: $10x - 12 + 3x = 10$

Simplify: $13x - 12 = 10$

Solve for: $x = \frac{22}{13}$

Use equation ① to find the y-coordinate of P:

$y = \frac{5}{2}x - 3$ where $x = \frac{22}{13}$ so $y = \frac{5}{2}\left(\frac{22}{13}\right) - 3$

$$= \frac{16}{13}$$

The coordinates of P are $\left(\frac{22}{13}, \frac{16}{13}\right)$.

> **Handy hint**
>
> This is called the method of **substitution**.

> **Handy hint**
>
> $\frac{22}{13}$ is the x-coordinate of P.

> **Handy hint**
>
> You could instead use equation ② to find the y-coordinate of P but this would be less straightforward.

> **Checkpoint**
>
> Check for yourself that the pair of values $x = \frac{22}{13}$, $y = \frac{16}{13}$ satisfy equation ②.

A Level Example 8

Line L_1 has equation $2y - x = 7$.

Line L_2 has equation $5y + 4x = 50$.

Find the area of the triangle formed by these two lines and the y-axis.

Working

You need to make a sketch to visualise the problem.

Line L_1 crosses the y-axis at $(0, 3.5)$ and the x-axis at $(-7, 0)$.

Line L_2 crosses the y-axis at $(0, 10)$ and the x-axis at $(12.5, 0)$.

Figure 1 shows a sketch of L_1 and L_2 which intersect at point P. Also shown are the y-intercepts of these lines.

Handy hint

Use the cover-up method – see Section 3.1.

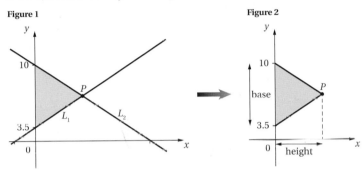

Figure 1

Figure 2

You need to find the area of the shaded triangle.

Use the vertical side for the base of this triangle, which has length $10 - 3.5 = 6.5$ (see Figure 2).

The (perpendicular) height of this triangle is then given by the x-coordinate of P.

To find the x-coordinate of P use elimination to solve the simultaneous equations:

$$2y - x = 7 \qquad ①$$
$$5y + 4x = 50 \qquad ②$$

Multiply ① by 4: $8y - 4x = 28 \qquad ③$

Add ② and ③: $13y = 78$

$$\text{so } y = \frac{78}{13} = 6$$

Use ① to find the x-coordinate of P.

When $y = 6$, the equation $2y - x = 7$ becomes $2(6) - x = 7$.

So: $\qquad\qquad 12 - x = 7$

Solve for x: $\qquad\qquad x = 5$

So the shaded triangle has base 6.5 and height 5.

The area of this triangle is $\frac{1}{2} \times 6.5 \times 5 = 16.25$ square units.

Handy hint

Turn the page 90° anticlockwise to visualize this base and height.

Handy hint

You could also use the method of substitution to solve these equations but elimination is easier in this case.

Handy hint

Use whichever equation is easiest for finding x, in this case, the equation $2y - x = 7$.

Common error

The solution to the equation $12 - x = 7$ is **not** 19.

⟫ Taking it further ⟫⟫

In A Level Maths, you will need to be able to find the intersection points of a curve and a line, or even a pair of curves. The method of substitution can be used to do this.

4 Algebra 2

4.1 Solving a quadratic equation by factorising

What you should already know:

- how to factorise a quadratic expression such as $x^2 - 5x + 6$.
- how to solve a quadratic equation such as $2x^2 + 9x + 4 = 0$ by factorising.

In this section you will learn:

- how to form and solve a quadratic equation as part of a more complex question.

Revisiting GCSE

At GCSE, you saw that factorising helps you to solve a quadratic equation.

GCSE Example 1

Solve the equation $3x^2 - 7x - 6 = 0$.

Working

There are several methods you could use to solve this equation but factorisation is generally the fastest.

Here is one way to factorise the expression $3x^2 - 7x - 6$.

Multiply the first and last numbers: $3 \times -6 = -18$

Find two numbers with product -18 and which combine to make the middle number -7 in the expression:

$$\text{Use } -9 \text{ and } 2, \text{ because } -9 \times 2 = -18$$
$$\text{and } -9 + 2 = -7.$$

So $3x^2 - 7x - 6 = \underline{3x^2 - 9x} + \underline{2x - 6}$.

> **Handy hint**
>
> Use the factorisation method you learnt at GCSE.

> **Handy hint**
>
> Write $-7x$ as $-9x + 2x$ rather than $2x - 9x$ to avoid issues with negative signs.

Factorise each group of terms: $= 3x(x - 3) + 2(x - 3)$

Fully factorise the expression: $= (3x + 2)(x - 3)$

The equation $3x^2 - 7x - 6 = 0$ can be written as
$$(3x + 2)(x - 3) = 0$$

So either $3x + 2 = 0$ or $x - 3 = 0$.

Solve for x: $x = -\dfrac{2}{3}$ or $x = 3$

›› Moving on to A Level ››››

In A Level questions, you often need to form and solve a quadratic equation as you work towards the solution of a more complicated problem.

A Level Example 2

The diagram shows a square lawn $ABCD$, with side length x metres, and a rectangular path $EFGH$, where $EF = (3x - 4)$ metres and $FG = (x \quad 2)$ metres.

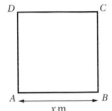

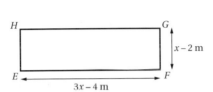

The lawn and the path have equal areas.

Find the perimeter of the path.

Working

To find the perimeter of the path, you first need to find the value of x.

Working in metres, the area of the square lawn is x^2 and the area of the rectangular path is $(3x - 4)(x \quad 2)$.

The two areas are equal, so: $(3x - 4)(x - 2) = x^2$

Expand the brackets: $3x^2 - 6x - 4x + 8 = x^2$

Simplify: $3x^2 - 10x + 8 = x^2$

Subtract x^2 from each side: $2x^2 - 10x + 8 = 0$

Divide each term by 2 $x^2 - 5x + 4 = 0$
to simplify the factorisation:

Factorise and solve $(x - 1)(x - 4) = 0$
this quadratic:

So $x = 1$ or $x = 4$.

$x = 1$ is not possible because then each side of the rectangle would have a negative length (for example, $EF = 3(1) - 4 = -1$).

So $x = 4$, $EF = 3(4) - 4$ and $FG = 4 - 2$
$= 8$ $= 2$

$EFGH$ is an 8 m by 2 m rectangle and so has perimeter 20 metres.

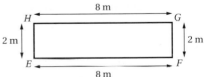

Quadratic equations feature in the whole of A Level Maths. They often arise when working through a complex problem and so it is important that you are able to factorise quickly and accurately.

4.2 Using the quadratic formula

What you should already know:

- how to use the quadratic formula to solve a quadratic equation such as $2x^2 - 3x - 1 = 0$.

In this section you will learn:

- how to use the quadratic formula with unknown coefficients.

>> **Revisiting GCSE** >> >>

At GCSE, you may have used the quadratic formula to solve an equation which could not easily be factorised. Exact answers obtained using the formula usually involve surds.

>> **Key point** >>

The quadratic equation $ax^2 + bx + c = 0$ has solutions

$$x = \frac{-b \pm \sqrt{b^2 - 4ac}}{2a}, \text{ where } a, b \text{ and } c \text{ are constants.}$$

Handy hint

The formula must only be applied to an equation of the form $ax^2 + bx + c = 0$ (that is, where one side is zero).

GCSE Example 3

Solve the equation $x^2 - 2x - 11 = 0$.

Give your answers in the form $a \pm b\sqrt{3}$ where a and b are positive integers.

Working

Apply the formula to the equation $x^2 - 2x - 11 = 0$, where $a = 1$, $b = -2$ and $c = -11$.

So: $x = \dfrac{-b \pm \sqrt{b^2 - 4ac}}{2a}$

$= \dfrac{-(-2) \pm \sqrt{(-2)^2 - 4(1)(-11)}}{2(1)}$

$= \dfrac{2 \pm \sqrt{4 - (-44)}}{2}$

$= \dfrac{2 \pm \sqrt{4 + 44}}{2}$

$= \dfrac{2 \pm \sqrt{48}}{2}$

$= \dfrac{2 \pm 4\sqrt{3}}{2}$

$= \dfrac{2}{2} \pm \dfrac{4\sqrt{3}}{2}$

$= 1 \pm 2\sqrt{3}$

Handy hint

To simplify $\sqrt{48}$, look for the *largest* square factor of 48:

$48 = 16 \times 3$ so $\sqrt{48} = 4\sqrt{3}$.

Common error

$\dfrac{2 \pm 4\sqrt{3}}{2}$ does **not** simplify to

$1 \pm 4\sqrt{3}$.

At A Level, you refer to the solutions to the quadratic equation $ax^2 + bx + c = 0$ as its **roots**. You also learn more about the real number system.

The real numbers consist of all the fractions (the rationals) together with numbers that cannot be written as a fraction, such as $\sqrt{2}$ and π (the irrationals).

Notice that if a is any real number then $a^2 \geq 0$.

For example, $(-2)^2 = (-2) \times (-2)$

$= 4$, a positive number.

But, if $a = \sqrt{-4}$ then $a^2 = -4$ is *negative*, and so $\sqrt{-4}$ cannot be a real number!

$\sqrt{d}$ is not a real number if $d < 0$.

>> **Key point** >>

You can now see that not every quadratic equation has roots that are real numbers.

For example, the roots of the equation $x^2 - 2x + 5 = 0$ are:

$$x = \frac{-b \pm \sqrt{b^2 - 4ac}}{2a}$$

$$= \frac{-(-2) \pm \sqrt{(-2)^2 - 4(1)(5)}}{2(1)}$$

$$= \frac{2 \pm \sqrt{4 - 20}}{2}$$

$$= \frac{2 \pm \sqrt{-16}}{2} \quad \text{and} \quad \sqrt{-16} \text{ is } not \text{ a real number.}$$

This example shows that whether or not a quadratic equation $ax^2 + bx + c = 0$ has real roots depends on the value of $b^2 - 4ac$.

>> **Key point** >>

For the quadratic equation $ax^2 + bx + c = 0$, the number $d = b^2 - 4ac$ is called the **discriminant** of the equation.

If $d < 0$ the equation has no real roots.

If $d = 0$ the equation has one real root.

If $d > 0$ the equation has two real roots.

Handy hint
If $d = 0$ then $x = \dfrac{-b \pm \sqrt{0}}{2a}$

$$= -\frac{b}{2a}$$

is the only root of this equation.

Here is an A Level question that uses the discriminant of a quadratic.

A Level Example 4

The equation $4x^2 - 12x + k = 0$, where k is a constant, has one real root.

a Find the value of k.

b For this value of k, state the number of real roots to the equation
$4x^2 - 12x - k = 0$.

Working

a The equation $4x^2 - 12x + k = 0$ has one real root which means the discriminant d of this equation must be zero.

You need to find d in terms of k: $d = b^2 - 4ac$

Use $a = 4$, $b = -12$ and $c = k$: $= (-12)^2 - 4(4)k$

$$= 144 - 16k$$

Use $d = 0$: $\quad\quad\quad\quad\quad 144 - 16k = 0$

Add $16k$ to each side: $\quad\quad\quad 144 = 16k$

Divide each side by 16: $\quad\quad \dfrac{144}{16} = k$

So $k = 9$.

b $k = 9$ so $-k = -9$.

You need to find the discriminant of the equation $4x^2 - 12x - 9 = 0$.

$d = b^2 - 4ac$ where $a = 4$, $b = -12$ and $c = -9$

$\quad = (-12)^2 - 4(4)(-9)$

$\quad = 144 - (-144)$

$\quad = 288 > 0$

As this discriminant is positive, the equation $4x^2 - 12x - 9 = 0$ has two real roots.

> **Handy hint**
>
> Answer the question being asked! You do not need to actually find these roots.

Where this topic goes next:

In Coordinate Geometry, the discriminant of a quadratic is used to determine whether or not a given line is a tangent to a quadratic curve or circle.

4.3 Further equation solving

What you should already know:

- how to solve a pair of simultaneous equations such as $y = x + 3$ and $x^2 + y^2 = 5$.

In this section you will learn:

- how to solve more complicated types of simultaneous equations.
- how to solve certain types of cubic equations.

▶▶ Revisiting GCSE ▶▶

An equation such as $y - 2x = 3$ is linear because each variable is raised to the power of 1. By contrast, the equation $x^2 + 2y - x = 5$ is non-linear, because it contains a variable which has been raised to a power different from 1 (that is, x^2).

At GCSE, you may have used the method of substitution to solve, for example, the simultaneous equations $y = x + 2$ and $x^2 + y^2 = 10$.

The solution tells you where the line $y = x + 2$ intersects the circle $x^2 + y^2 = 10$.

Here is a GCSE-style question which uses these equations.

> **Handy hint**
>
> See Section 3.4 for the method of substitution.

GCSE Example 5

The diagram shows the circle $x^2 + y^2 = 10$ and the line $y = x + 2$.

The line and the circle intersect at the points A and B.

Find the coordinates of A and B.

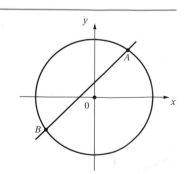

Working

Label the equations: $\quad\quad y = x + 2 \quad$ ①

$\quad\quad\quad\quad\quad\quad\quad x^2 + y^2 = 10 \quad$ ②

Use the method of substitution by replacing the y term in equation ② with the expression $(x + 2)$ from equation ①.

Equation ② becomes: $x^2 + (x + 2)^2 = 10$

Expand the bracket: $x^2 + x^2 + 4x + 4 = 10$

Simplify and rearrange
to make one side zero: $2x^2 + 4x - 6 = 0$

Divide all terms by 2: $x^2 + 2x - 3 = 0$

Factorise: $(x - 1)(x + 3) = 0$

So, either $x = 1$ or $x = -3$.

Substitute each of these answers for x into the line equation $y = x + 2$ to find the y-coordinates of A and B:

When $x = 1$, $y = x + 2$ When $x = -3$, $y = x + 2$

$= 1 + 2$ $= -3 + 2$

$= 3$ $= -1$

A has coordinates $(1, 3)$. B has coordinates $(-3, -1)$.

At A Level, you may be given a pair of simultaneous equations to solve where one equation is a mixture of squared and linear terms.

Handy hint

From the positions of A and B in the diagram, these answers show that the x-coordinate of A is 1 and the x-coordinate of B is -3.

Handy hint

Do **not** use equation ② to find these y-coordinates as this can produce answers which are not solutions to both equations.

A Level Example 6

Solve the simultaneous equations $x - 2y = 1$, $x^2 + 3x + y^2 = 4$.

Working

Label the equations: $x - 2y = 1$ ①

                          $x^2 + 3x + y^2 = 4$ ②

You need to rearrange equation ① to make one of its variables the subject.

It is easier to make x the subject of equation ① than y.

Equation ①: $x - 2y = 1$

Add $2y$ to each side: $x = 1 + 2y$

Substitute each x term in equation ② with the expression $(1 + 2y)$.

Equation ② becomes: $(1 + 2y)^2 + 3(1 + 2y) + y^2 = 4$

Expand all brackets: $1 + 4y + 4y^2 + 3 + 6y + y^2 = 4$

Simplify and re-arrange
to make one side zero: $5y^2 + 10y = 0$

Factorise and solve: $5y(y + 2) = 0$

So $y = 0$ or $y = -2$.

Use the rearrangement $x = 1 + 2y$ of equation ① to find the corresponding x values.

When $y = 0$, $x = 1 + 2y$ When $y = -2$, $x = 1 + 2y$

$= 1 + 2(0)$ $= 1 + 2(-2)$

$= 1$ $= -3$

So, the solutions are $(x = 1, y = 0)$, $(x = -3, y = -2)$.

At A Level, you will need to know how to solve certain types of cubic equations (that is, equations with highest power 3).

For example, $(x + 1)(x - 2)(x - 4) = 0$ is a cubic equation because, after expanding the brackets, the highest power of x in the equation is 3.

To solve the equation $(x + 1)(x - 2)(x - 4) = 0$

Set each bracket to zero: $(x + 1) = 0$ or $(x - 2) = 0$ or $(x - 4) = 0$

Handy hint

Try to think ahead here – if you were to rearrange $x - 2y = 1$ for y, the answer would involve fractions.

Common error

$(1 + 2y)^2$ is **not** $1 + 4y^2$. Write $(1 + 2y)^2$ as $(1 + 2y)(1 + 2y)$ to avoid this error.

Common error

To solve $5y^2 + 10y = 0$ do **not** divide through by y. If you do, you will lose the answer $y = 0$.

Handy hint

You can list the solutions in pairs. It is conventional to list x before y, even if the y values were found first.

Handy hint

This is the same method used when solving a quadratic equation in factorised form.

So the solutions (or roots) are: $x = -1, 2, 4$.

You may need to factorise a cubic before you can solve an equation.

Solve the equation $2x^3 - 11x^2 + 5x = 0$.

Working

You need to write the equation in factorised form:

$$2x^3 - 11x^2 + 5x = 0$$

Take out a factor of x: $\quad x(2x^2 - 11x + 5) = 0$

Factorise the quadratic: $\quad x(2x - 1)(x - 5) = 0$

Solve for x: $\qquad\qquad\qquad\qquad x = 0, \dfrac{1}{2}, 5$

Common error

Do **not** divide through by x. If you do, you will miss the solution $x = 0$.

Handy hint

You could write the factorised equation as $(x - \mathbf{0})(2x - 1)(x - 5) = 0$.

Where this topic goes next:

Solving linear and non-linear simultaneous equations enables you to find where a line intersects a curve. This skill is required in many Coordinate Geometry questions, particularly when working with lines and circles. You can sketch a cubic curve if you know how to find its roots.

5 Coordinate geometry 2

5.1 Transformations of graphs

What you should already know:

- how the graphs of $y = f(x) + a$, $y = f(x + a)$, $y = f(-x)$ and $y = -f(x)$, where a is a constant, are related to the graph of $y = f(x)$.

In this section you will learn:

- how to keep track of the important points on a curve as it undergoes a transformation.

>> **Revisiting GCSE** >>

At GCSE, you met the terms translation, stretch, reflection and rotation to describe transformations of shapes and graphs. For example, the translation vector $\begin{pmatrix} 2 \\ -3 \end{pmatrix}$ means that any point on a graph is shifted 2 units to the right along the x-axis and 3 units down the y-axis. So, using this vector, the point $P(1, 4)$ has *image* $P'(3, 1)$.

Handy hint

You add the vectors:
$\begin{pmatrix} 1 \\ 4 \end{pmatrix} + \begin{pmatrix} 2 \\ -3 \end{pmatrix} = \begin{pmatrix} 1 + 2 \\ 4 - 3 \end{pmatrix} = \begin{pmatrix} 3 \\ 1 \end{pmatrix}$

>> **Key point** >>

Here are some rules for transforming the graph of $y = f(x)$.

For a is a constant:

1 The graph of $y = f(x) + a$ is a translation of $y = f(x)$ by the vector $\begin{pmatrix} 0 \\ a \end{pmatrix}$.

2 The graph of $y = f(x + a)$ is a translation of $y = f(x)$ by the vector $\begin{pmatrix} -a \\ 0 \end{pmatrix}$.

3 The graph of $y = f(-x)$ is a reflection of $y = f(x)$ in the y-axis.

4 The graph of $y = -f(x)$ is a reflection of $y = f(x)$ in the x-axis.

GCSE Example 1

Diagram 1 shows the graph of $y = f(x)$. On one copy of the diagram, sketch the graph of:

a $y = f(x + 3)$

b $y = -f(x)$.

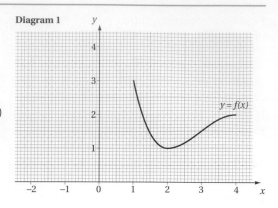

Diagram 1

Working

a   The graph of $y = f(x + 3)$ is a translation of $y = f(x)$ by the vector $\begin{pmatrix} -3 \\ 0 \end{pmatrix}$. See Graph A on Diagram 2.

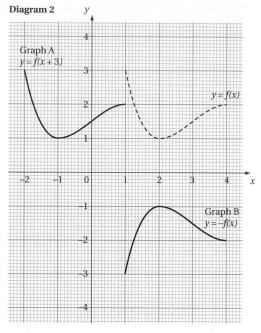

Diagram 2

> **Common error**
>
> $y = f(x + 3)$ is **not** a translation of $y = f(x)$ along the y-axis.

b   The graph of $y = -f(x)$ is a reflection of $y = f(x)$ in the x-axis. See Graph B on Diagram 2.

Moving on to A Level

At A Level, you will learn how a graph can be **stretched** in a direction which is parallel to either the y-axis or the x-axis. For example, the dotted curve shown in the diagram is the graph of $y = \sin x$ for $0° \le x \le 360°$. Also shown is the graph of $y = 2\sin x$, which is a stretch of $y = \sin x$, scale factor 2 from the origin, O, parallel to the y-axis.

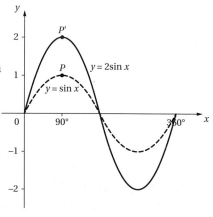

Key point

Here are the rules for stretching the graph of $y = f(x)$ from the origin O.

For a is a positive constant:

5   The graph of $y = af(x)$ is a stretch of $y = f(x)$ parallel to the y-axis, scale factor a.

6   The graph of $y = f(ax)$ is a stretch of $y = f(x)$ parallel to the x-axis, scale factor $\dfrac{1}{a}$.

A stretch scale factor $\dfrac{1}{2}$, for example, is the mathematical description for compression of a graph by scale factor 2.

You may need to keep track of specific points on a graph as it undergoes a transformation.

> *Handy hint*
>
> When $x = 90°$ the graph of $y = \sin x$ reaches its maximum value 1 and the graph of $y = 2\sin x$ reaches its maximum value 2.
>
> $P(90°, 1) \xrightarrow{\; 2 \times (y\,\text{co-ordinate}) \;} P'(90°, 2)$

A Level Example 2

Figure 1 shows the graph of $y = f(x)$. The graph has a maximum point at $A(3, 2)$ and a minimum point at $C(0, 1)$. The graph crosses the x-axis at the point $B(6, 0)$.

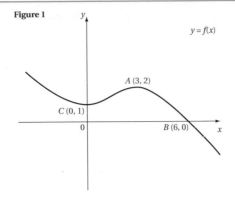

Figure 1

Sketch, on separate diagrams, the curve with equation:

a $y = 2f(x)$

b $y = f(3x)$.

On each diagram show clearly the coordinates of the maximum and minimum points and where the curve crosses the x-axis.

A Level Alert!

You need to keep track of three points under each of these transformations. GCSE questions do not normally ask for this level of tracking.

Working

a You can sketch the graph of $y = 2f(x)$ by stretching the graph of $y = f(x)$ scale factor 2 parallel to the y-axis from the origin O.

This means every point on the graph of $y = 2f(x)$ is the result of doubling the y-coordinate of the corresponding point on the graph of $y = f(x)$.

Applying this transformation to the coordinates of the points A, B and C gives:

$$A(3, 2) \xrightarrow{\ 2 \times (y\text{-coordinate})\ } A'(3, 4)$$

$$B(6, 0) \xrightarrow{\ 2 \times (y\text{-coordinate})\ } B'(6, 0)$$

$$C(0, 1) \xrightarrow{\ 2 \times (y\text{-coordinate})\ } C'(0, 2)$$

The transformed graph is shown on Figure 2.

Figure 2

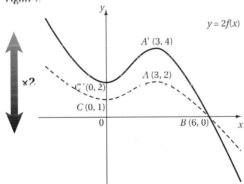

Handy hint

Under this transformation, points on the x-axis do not move. So $B' = B$.

Handy hint

You can sketch the dotted graph $y = f(x)$ to help visualise the stretch, but your final answer should just show the graph $y = 2f(x)$.

Common error

The transformed graph is **not** a translation of $y = f(x)$ along the y-axis.

Although there is some movement along the y-axis (for example, $C \rightarrow C'$), the transformed graph has a different shape to that of $y = f(x)$.

A translation does not alter the shape of a graph.

b You must start with the graph $y = f(x)$.

The graph of $y = f(3x)$ is a stretch parallel to the x-axis of the graph $y = f(x)$, scale factor $\frac{1}{3}$ from the origin O.

This means every point on the graph of $y = f(3x)$ is the result of dividing the x-coordinate of the corresponding point on the graph of $y = f(x)$.

Common error

$y = f(3x)$ is **not** a stretch scale factor 3 along the x-axis.

Applying this transformation to the coordinates of the points A, B and C gives:

$$A(3, 2) \xrightarrow{\text{(x-coordinate)} \div 3} A'(1, 2)$$

$$B(6, 0) \xrightarrow{\text{(x-coordinate)} \div 3} B'(2, 0)$$

$$C(0, 1) \xrightarrow{\text{(x-coordinate)} \div 3} C'(0, 1).$$

The transformed graph is shown in Figure 3.

>>> **Taking it further** >>>>>

Transformations are used throughout A Level Maths. For example, it is possible to describe the equation of a circle with centre at the point (3, 4) as a transformation of a circle centred at the origin. You will also meet transformations applied to trigonometric graphs to help describe the behaviour of oscillating systems.

> **Handy hint**
>
> Under this transformation, points on the y-axis do not move. So $C' = C$.

Figure 3

5.2 Sketching curves

What you should already know:

- how to find the turning point on a quadratic curve when its equation is expressed in completed square form.
- how to sketch simple cubic curves such as the graphs of $y = x^3 - 1$ and $y = -x^3$.

In this section you will learn:

- how to sketch a quadratic curve when its equation is in factorised form.
- how to sketch more complicated cubic curves such as the graph of $y = x^3 - x$.

At GCSE, you learnt how to find the coordinates of a turning point (the minimum or maximum point) of a quadratic curve by first expressing its equation in completed square form.

GCSE Example 3

a Express $x^2 + 4x + 5$ in the form $(x + p)^2 + q$ where p and q are integers.

b Sketch the curve with equation $y = x^2 + 4x + 5$, labelling the minimum point with its coordinates.

> **Handy hint**
>
> Use the method you learnt at GCSE to complete the square.

Working

a Here is one method for writing $x^2 + 4x + 5$ as a completed square.

Separate the constant term 5 from the other terms:
$$x^2 + 4x \qquad + 5$$
Complete the square on $x^2 + 4x$:
$$(x + 2)^2 - 2^2 \qquad + 5$$
Simplify: $\qquad (x + 2)^2 + 1$
So $x^2 + 4x + 5 = (x + 2)^2 + 1$.

> **Checkpoint**
>
> $y = (x + 2)^2 + 1$ is a translation of $y = x^2$ through the vector $\begin{pmatrix} -2 \\ 1 \end{pmatrix}$
> – see Section 5.1

b Since $(x + 2)^2 \geq 0$ for all real values of x, the minimum value of $(x + 2)^2 + 1$ is 1 which occurs when $x = -2$.
The minimum point on the curve with equation $y = x^2 + 4x + 5$ has coordinates $(-2, 1)$.
Here is a sketch of this curve.

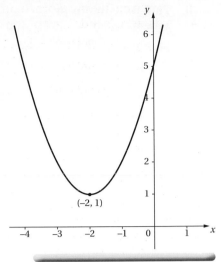

Key point

The quadratic curve with equation:

$$y = (x + p)^2 + q$$

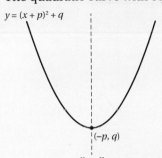

$(-p, q)$

$x = -p$

Handy hint

If you reflected this curve in the line $x = -p$, the diagram would look the same.

Handy hint

Label the y-intercept with its value. When $x = 0$, $y = (0)^2 + 2(0) + 5 = 5$.

- $y = (x + p)^2 + q$ has minimum turning point at $(-p, q)$.
- $y = -(x + p)^2 + q$ has maximum turning point at $(-p, q)$ where p and q are constants.

In either case, the curve is symmetrical in the vertical line $x = -p$.

At A Level, you may be asked to sketch a quadratic curve by expressing its equation in factorised form. This is sometimes easier than writing the equation as a completed square.

A Level Example 4

Sketch, on separate diagrams, the graphs with these equations.

a $y = 2x^2 - x - 6$

b $y = -x^2 + 4x - 4$

Handy hint

Use any appropriate method to factorise – see Section 4.1.

Working

a Find the y-intercept:
When $x = 0$,
$y = 2(0)^2 - (0) - 6 = -6$.

Find the x-intercepts:
When $y = 0$,
$2x^2 - x - 6 = 0$.

Solve this equation by factorisation:

$$(2x + 3)(x - 2) = 0$$

So $x = -\dfrac{3}{2}$ or $x = 2$.

$x = -\dfrac{3}{2}$ and $x = 2$ are the **roots** of this graph which means the graph crosses the x-axis at $x = -\dfrac{3}{2}$ and $x = 2$.

Here is a sketch of this curve.

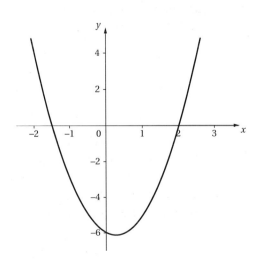

b Find the y-intercept:

When $x = 0$, $y = -(0)^2 + 4(0) - 4 = -4$.

Find the roots:

When $y = 0$, $-x^2 + 4x - 4 = 0$.

Multiply through by (-1)
to make factorising the equation: $x^2 - 4x + 4 = 0$

Factorise: $(x - 2)^2 = 0$

So $x = 2$ is the only real root of this equation which means the curve touches, but does not cross, the x-axis at $x = 2$.

Here is a sketch of this curve.

Handy hint

The roots of an equation are the same as the roots of its graph - see Section 4.2.

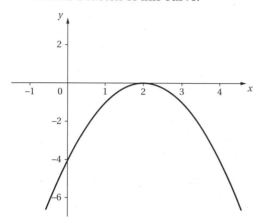

Checkpoint

The graph of $y = -x^2 + 4x - 4$ must be $\cap$-shaped because the coefficient of x^2 is -1 (that is, negative).

The y-intercept is -4 so the graph is $\cap$-shaped.

You can use the same method to sketch the graph of the cubic equation $y = (x - a)(x - b)(x - c)$.

A Level Example 5

Sketch the graph with equation $y = (x + 2)(x - 1)(x - 3)$.

Working

Find the y-intercept:

When $x = 0$, $y = (0 + 2)(0 - 1)(0 - 3)$
$$= (2)(-1)(-3)$$
$$= 6$$

Find the roots:

When $y = 0$, $(x + 2)(x - 1)(x - 3) = 0$.

So either $x = -2$, $x = 1$ or $x = 3$.

Handy hint

There is no point in multiplying out these brackets in order to sketch the graph.

Handy hint

See Section 4.3 for solving a cubic equation.

Here is a sketch of this curve.

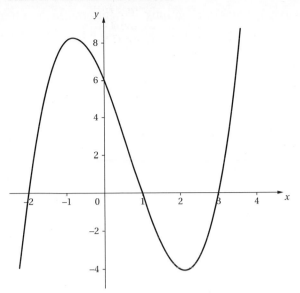

A Level Alert!

You may be asked to find the coordinates of the turning points on this curve – see Chapter 8 Differentiation.

Key point

If $(x - a)$ is a factor of $f(x)$ then $x = a$ is a root of the graph with equation $y = f(x)$.

If the same factor appears more than once in the equation, then the graph touches but does not cross the x-axis at the corresponding root.

A Level Example 6

Sketch the graph with equation $y = x^3 - 6x^2 + 9x$.

Working

Factorise the equation of the curve as far as possible.

$$y = x^3 - 6x^2 + 9x$$
$$= x(x^2 - 6x + 9)$$
$$= x\,(x - 3)(x - 3)$$

So the roots of this graph are $x = 0$ and $x = 3$ (repeated).

The graph passes through the origin and touches, but does not cross, the x-axis at $x = 3$.

The coefficient of x^3 in the equation is $+1$ and so the curve is $\checkmark$-shaped.

Here is a sketch of this graph.

Handy hint

$x = 3$ is called a **repeated** root, and $(x - 3)$ a **repeated** factor, of this equation.

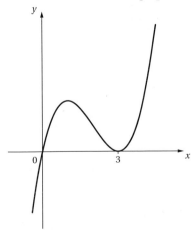

Where this topic goes next:

At A Level, you will use combined transformations to help you sketch graphs of trigonometric functions, such as $y = 2 \sin(x + 30°)$, exponential functions, such as $y = 3 - 2^{-x}$ and reciprocal functions, such as $y = \frac{2}{x-1} + 3$.

5.3 Intersection points of graphs

What you should already know:

* how to find approximate solutions to an equation using a graphical method.
* how to show an equation does not have any solutions using a graphical method.

In this section you will learn:

* how to use algebra to solve geometrical problems involving lines and curves.

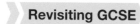 **Revisiting GCSE**

At GCSE, you may have found an approximate solution to an equation by using a graphical method.

GCSE Example 7

The diagram shows the curve $y = x^2 - 4x + 7$ and the line $y = -x + 6$, which intersect at points P and Q. The minimum point on the curve is at $(2, 3)$.

Use the diagram to:

a find approximate solutions to the equation $x^2 - 3x + 1 = 0$

b explain why the equation $x^2 - 4x + 5 = 0$ has no real solutions.

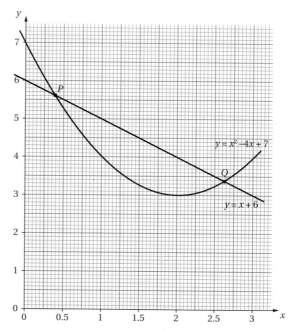

Working

a You need to compare the given equation $y = x^2 - 4x + 7$ with the new equation $x^2 - 3x + 1 = 0$.

Write down the given equation:
$y = x^2 - 4x + 7$ ①

Write down the new equation in reverse:
$0 = x^2 - 3x + 1$ ②

Subtract these equations:
① − ② $y = -x + 6$ ③

Equation ③ is the line drawn in the diagram.

So the solutions to the equation $x^2 - 3x + 1 = 0$ are the x-coordinates of points P and Q shown in the diagram, which are approximately 0.4 and 2.6.

So the equation $x^2 - 3x + 1 = 0$ has approximate solutions 0.4 and 2.6.

> **Common error**
> Equation ③ is **not** $y = -7x + 6$.
> Take care with signs: $-4x - (-3x) = -4x + 3x = -x$.

b Using a similar approach to that in part **a**.

Write down the given equation:
$y = x^2 - 4x + 7$ ①

Write down the new equation in reverse:
$0 = x^2 - 4x + 5$ ②

Subtract these equations: ① − ② $y = 2$

The line $y = 2$ has been added to the first diagram.

In the diagram, the minimum point of the curve has y-coordinate 3, so the line $y = 2$ does not intersect this curve. Hence the equation $x^2 - 4x + 5 = 0$ has no real solutions.

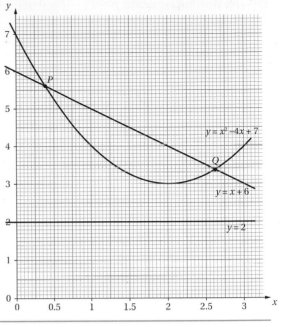

At A level, approximate solutions are not very useful. You will usually be asked to find the exact coordinates of the points where two graphs intersect.

A Level Example 8

The diagram shows the line with equation $y - 2x - 4 = 0$. Point A is the y-intercept of this line.

Also shown is the curve with equation $y = x^2 - 6x + 16$.

This line and curve intersect at points B and C.

Find the value of the constant k such that $AB = k \times BC$.

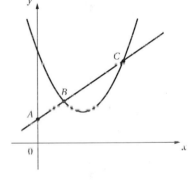

Working

You need to find the coordinates of points A, B and C.

Find the y-intercept of the line.

When $x = 0$, $y - 2(0) - 4 = 0$

So: $y - 4$

So A has coordinates $(0, 4)$.

Points B and C are where the line and curve intersect. You can find the coordinates of these points by solving simultaneous equations using the method of substitution.

$$y - 2x - 4 = 0 \quad ①$$
$$y = x^2 - 6x + 16 \quad ②$$

Replace the y term in equation ① with the expression $x^2 - 6x + 16$ from equation ②.

Equation ① becomes: $(x^2 - 6x + 16) - 2x - 4 = 0$

Simplify: $x^2 - 8x + 12 = 0$

Factorise: $(x - 2)(x - 6) = 0$

So either $x = 2$ or $x = 6$.

Handy hint

See Section 3.4.

Handy hint

From the diagram you can see that the x-coordinate of B is less than the x-coordinate of C.

Use the line equation $y - 2x - 4 = 0$ to find the y-coordinates of B and C.

When $x = 2$, $y - 2(2) - 4 = 0$ When $x = 6$, $y - 2(6) - 4 = 0$

So: $y = 8$ So: $y = 16$

Checkpoint

Use the curve equation to check $B(2, 8)$ and $C(6, 16)$ lie on this curve.

So these points have coordinates $A(0, 4)$, $B(2, 8)$ and $C(6, 16)$.

You can now use the distance formula to find the length of AB and BC.

$AB = \sqrt{(2-0)^2 + (8-4)^2}$ and $BC = \sqrt{(6-2)^2 + (16-8)^2}$

$\quad = \sqrt{20}$ $\qquad\qquad\qquad\quad = \sqrt{80}$

$\quad = 2\sqrt{5}$ $\qquad\qquad\qquad\quad = 4\sqrt{5}$

Handy hint

See Section 3.3.

Comparing these answers, $AB = \dfrac{1}{2} \times BC$, so $k = \dfrac{1}{2}$.

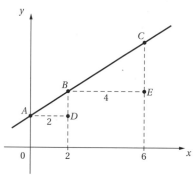

A Level Alert!

This problem:
- tests different areas of maths
- is not broken down into steps
- is solvable using different methods.

This is typical of an A Level question.

You could also have found k by noticing that triangles ABD and BCE are similar.

Since $AD = \dfrac{1}{2} \times BE$ it follows that $AB = \dfrac{1}{2} \times BC$.

Key point

To find the x-coordinates of the points where a line and curve intersect, solve a pair of simultaneous equations.

If no solutions to these equations exist, the line and curve do not intersect.

If there is exactly one solution to these equations, the line is a **tangent** to the curve.

Taking it further

Finding intersection points is often a required first step when calculating an area of a region using integration.

6 Trigonometry

6.1 Trigonometry and triangles

What you should already know:

- how to use the sine and cosine rules to find angles and lengths of sides of a triangle.
- how to find the area of a sector of a circle.

In this section you will learn:

- how to apply these rules to solve problems set in a context.

❯❯ Revisiting GCSE ❯❯

At GCSE, you learnt how to use trigonometric rules (usually called SOHCAHTOA) for right-angled triangles. You also met the sine and cosine rules, which can be applied to any triangle (including right-angled triangles!).

❯ Key point

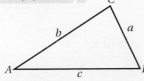

ABC is any triangle where $AB = c$, $AC = b$, $BC = a$ and $\hat{A}$ is the angle at vertex A, and so on.

Sine rule: $\dfrac{a}{\sin \hat{A}} = \dfrac{b}{\sin \hat{B}} = \dfrac{c}{\sin \hat{C}}$.

By inverting each fraction, the sine rule can be written as
$$\frac{\sin \hat{A}}{a} = \frac{\sin \hat{B}}{b} = \frac{\sin \hat{C}}{c}.$$

Cosine rule: $a^2 = b^2 + c^2 - 2bc \cos \hat{A}$.

Which can be rearranged to give $\cos \hat{A} = \dfrac{b^2 + c^2 - a^2}{2bc}$.

Handy hint

The rule $\dfrac{\sin \hat{A}}{a} = \dfrac{\sin \hat{B}}{b} = \dfrac{\sin \hat{C}}{c}$ is useful for finding angles.

Handy hint

There are three cosine rules but they all have the same form. Write down the other two versions.

Handy hint

These formulae are **not** given to you in the exam. You have to remember them.

You may need to use the sine and cosine rules in the same question.

GCSE Example 1

In triangle ABC, point D on BC is such that $BD = 11$ cm, $DC = 6$ cm, angle $DAC = 25°$ and angle $DCA = 35°$.

Find the perimeter of triangle ABD.

Give the answer to 1 decimal place.

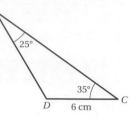

Working

You need to find the lengths AD and AB.

You can use the sine rule on triangle ADC to find the length AD.

Sine rule: $\dfrac{AD}{\sin 35°} = \dfrac{6}{\sin 25°}$

Multiply both sides by $\sin 35°$: $AD = \left(\dfrac{6}{\sin 25°}\right) \times \sin 35°$

$= 8.14318…$

So $AD = 8.14$ cm (2 decimal places).

In triangle ADC, angle $ADC = 180° - (25° + 35°)$

$= 180° - 60°$

$= 120°$

So in triangle ABD, $\angle ADB = 180° - 120°$

$= 60°$

Now use the cosine rule on triangle ABD to find the length AB.

Cosine rule: $AB^2 = 11^2 + 8.14^2 - [2(11)(8.14)\cos 60°]$

$= 97.7196$

So $AB = \sqrt{97.7196}$

$= 9.8853…$

$= 9.89$ cm (2 decimal places)

Working to 2 decimal places,
the perimeter of triangle $ABD = 11 + 8.14 + 9.89$

$= 29.03$

$= 29.0$ cm (1 decimal place)

At A Level, a trigonometry question can be set in a real-life context.

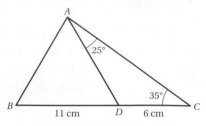

Handy hint

Use the version of the sine rule which has side lengths on the numerators.

Handy hint

The vertices are not always A, B and C – you have to adapt the rule to fit each particular triangle.

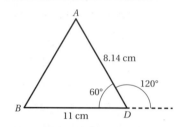

Common error

AD is **not** $\dfrac{\sin (35° \times 6)}{\sin 25°}$.

Handy hint

Round intermediate values to 2 d.p. to obtain a final answer accurate to 1 d.p.

So use 8.14 for the length of AD.

Common error

The answer is **not** 29 cm to 1 decimal place, even though 29 and 29.0 are numerically equal.

A Level Example 2

The diagram shows the plan of a recreation park. ABC is in the shape of a sector of a circle, centre B and radius 1.5 km.

The dotted line AC represents a path.

Starting at A, a man walks along the path AC in 30 minutes.

Without stopping, he then walks the perimeter of the park.

This is the end of his walk.

His speed at all times is 5 km per hour.

Find the total time it takes him to complete his walk.

Give your answer in hours and minutes, to the nearest minute.

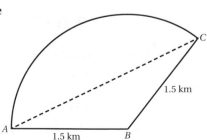

Working

The perimeter of the park includes the arc AC. To find this arc length, first calculate the length of AC and then calculate the size of angle ABC.

Start by using the given information:

Time taken to walk the path AC = 0.5 hours

Walking speed = 5 km/hour

So length of path AC = 0.5 × 5

$= 2.5$ km

Use the rearranged cosine rule to find angle ABC.

$$\cos \hat{B} = \frac{1.5^2 + 1.5^2 - 2.5^2}{2(1.5)(1.5)}$$

$$= -\frac{7}{18}$$

So $\hat{B} = \cos^{-1}\left(-\frac{7}{18}\right)$

$= 112.88...°$

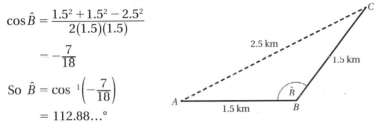

Working in km, the arc length $AC = \left(\dfrac{112.88...°}{360°}\right) \times 2\pi \times 1.5$

$= 2.955...$ km

The perimeter of the park is $1.5 + 1.5 + 2.955... = 5.955...$ km. With average speed 5 km/hour, the time taken to walk the perimeter of the park is $\dfrac{5.955...}{5} = 1.191...$ hours.

Total walking time = $(1.191... + 0.5)$ hours

$= 1.691...$ hours

$= 1$ hour 41 minutes (nearest minute)

Key point

In general, if you know three sides of a triangle, or two sides and the angle between them, use the cosine rule. Otherwise use the sine rule.

Where this topic goes next:

You will need to use the sine and cosine rules to answer questions involving circle geometry. You will be expected to decide for yourself which of these rules you need to use when solving a problem.

6.2 The area of any triangle

What you should already know:

- how to find the area of any triangle using two sides and the angle between them.

In this section you will learn:

- how to apply the area formula to a triangle which is part of a compound shape.

Revisiting GCSE

At GCSE, you used the formula $Area = \frac{1}{2}\, base \times height$ to find the area of a right-angled triangle. You may also have met a more general formula which finds the area of **any** triangle.

Key point

ABC is any triangle. $AB = c$, $AC = b$, $BC = a$.

Area of triangle ABC $= \frac{1}{2}ab \sin \hat{C}$, where $\hat{C}$ is the angle at vertex C.

(You can also write Area $= \frac{1}{2}ac \sin \hat{B}$ and Area $= \frac{1}{2}bc \sin \hat{A}$.)

You use this formula if you know the length of two sides **and** the angle between them.

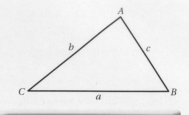

Handy hint

You need to remember the area formula – it won't be given to you in the exam.

GCSE Example 3

In triangle ABC, $AB = x$, $AC = x + 2$, $BC = x + 3$ where all lengths are in cm. Angle $ABC = 60°$.

The perimeter of triangle ABC is 20 cm.

Find the exact area of triangle ABC.

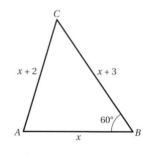

Working

Using the given information, you need to find the value of x and then use the formula Area $= \frac{1}{2}ac \sin \hat{B}$.

The perimeter P of this triangle is given by:
$$P = x + (x + 2) + (x + 3)$$
$$= 3x + 5$$

The numerical value of P is 20, so solve $3x + 5 = 20$.

Subtract 5 from each side: $\qquad\qquad 3x = 15$

Divide each side by 3: $\qquad\qquad x = 5$

So $AB = 5$ cm and $BC = 5 + 3 = 8$ cm.

You are given angle $\hat{B}$ so Area $= \frac{1}{2}ac \sin \hat{B}$

$$= \frac{1}{2}(8)(5) \sin 60°$$
$$= 10\sqrt{3} \text{ cm}^2$$

Common error

The exact answer is **not** 17.320 508 08.

No surd can be described in decimal form.

At A Level you may be given a compound shape, one of whose parts is a triangle.

A Level Example 4

The diagram shows a metal plate in which triangle ABC has been welded onto a semi-circular section. The semi-circle has diameter BC.

$AB = 12$ cm, $AC = 12$ cm and angle $BCA = 40°$.

Find, to the nearest square cm, the area of the plate.

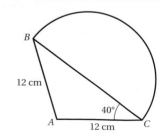

Working

You need to find the area of triangle ABC.

$AB = AC$ which means triangle ABC is isosceles.

So angle $ABC = 40°$ and hence angle BAC $= 180° - (40° + 40°)$

$= 100°$

Area of triangle ABC $= \frac{1}{2}bc \sin \hat{A}$

$= \frac{1}{2}(12)(12) \sin 100°$

$= 70.906... \text{ cm}^2$

You can find the area of the semi-circle by finding the length of the diameter BC.

Use the sine rule on triangle ABC to find side $BC (= a)$.

$\frac{a}{\sin 100°} = \frac{12}{\sin 40°}$ so $a = \left(\frac{12}{\sin 40°}\right) \times \sin 100°$

$= 18.385...$

The semi-circle has radius $r = \frac{18.385...}{2} = 9.192...\text{cm}$.

So the area of the semi-circle is $\frac{1}{2}\pi r^2 = \frac{1}{2}\pi(9.192...)^2$

$= 132.736... \text{ cm}^2$

Hence, the area of the plate $= 70.906... + 132.736...$

$= 203.642... \text{ cm}^2$

$= 204 \text{ cm}^2$ (nearest square cm)

Where this topic goes next:

You will need to use the formula for the area of any triangle to solve various geometrical problems, including finding the area of a segment of a circle.

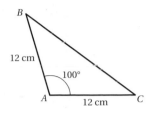

6.3 Solving a trigonometric equation

What you should already know:

- the general shape and properties of the sine and cosine curves.
- how to find another solution to a simple trigonometric equation by using a given solution.

In this section you will learn:

- how to find all the solutions to a more complex trigonometric equation over a given range.

Revisiting GCSE

At GCSE, you may have been given a solution to an equation involving sine or cosine and asked to find another solution to the same equation. One way to do this is to use the graphs of $y = \sin x$ or $y = \cos x$.

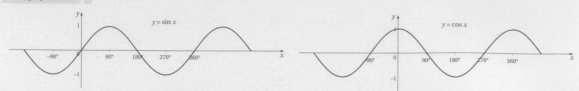

These graphs:

1   are cyclic with period 360°

2   have a maximum value of 1 and a minimum value of -1

3   are defined for any positive or negative value of x.

Some important values on the sine graph are:

$\sin 0° = \sin 180° = \sin 360° = 0$, $\sin 90° = 1$, $\sin 270° = -1$.

Some important values on the cosine graph are:

$\cos 0° = \cos 360° = 1$, $\cos 90° = \cos 270° = 0$, $\cos 180° = -1$.

Handy hint

'period 360°' means $\sin(x° + 360°) = \sin x°$ for all values of x and that 360° is the *smallest* angle with this property.

GCSE Example 5

The diagram shows the graph of $y = \cos x$ for $0° \leqslant x \leqslant 360°$.

a   Write down the solutions of the equation $\cos x = 0$ for $0° \leqslant x \leqslant 360°$.

One solution of the equation $\cos x = 0.5$ is $x = 60°$.

b   Find another solution of the equation $\cos x = 0.5$ for $0° \leqslant x \leqslant 360°$.

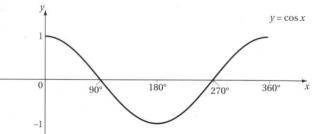

Working

a   You can see from the graph that $\cos x = 0$ where the graph of $y = \cos x$ crosses the x-axis.

The solutions of $\cos x = 0$ for $0° \leqslant x \leqslant 360°$ are $x = 90°$ or $x = 270°$.

Handy hint

You must find all the values of x between 0° and 360° for which $\cos x = 0$.

b   One solution of the equation $\cos x = 0.5$ is $x = 60°$.

This means $\cos 60° = 0.5$.

So the line $y = 0.5$ intersects the graph of $y = \cos x$ at the point where $x = 60°$.

This intersection point is shown in Figure 1.

Handy hint

The line with equation $y = 0.5$ is a horizontal line with y-intercept 0.5.

Intersection points correspond to solutions to equations – see 5.3.

Figure 1

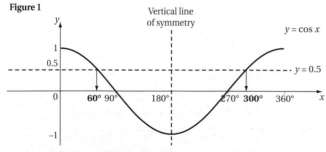

The next positive value of x where the line $y = 0.5$ intersects the graph $y = \cos x$ is where $x = 360° - 60°$

$$= 300°$$

So $x = 300°$ is the other solution of the equation $\cos x = 0.5$ in the range $0° \leqslant x \leqslant 360°$.

Checkpoint

This curve is symmetrical in the vertical line $x = 180°$.

So $x = 360° - 60°$
$= 300°$
is another solution of this equation.

Checkpoint

Use your calculator to check that $\cos 300° = 0.5$.

Moving on to A Level

At A Level, you will need to be able to solve more complicated equations without first being given a solution to work with.

A Level Example 6

Solve the equation $3 \sin x + 2 = 3$ for $0° \leqslant x \leqslant 360°$. Give each answer to 1 decimal place.

Working

You first need to find the value of $\sin x$.

Given equation: $\qquad 3 \sin x + 2 = 3$

Subtract 2 from both sides: $\quad 3 \sin x = 1$

Divide both sides by 3: $\qquad \sin x = \dfrac{1}{3}$

Find one solution by using $\boxed{\sin^{-1}}$: $x = \sin^{-1}\left(\dfrac{1}{3}\right)$
$$= 19.471...$$
$$= 19.5° \text{ (1 decimal place)}.$$

Now sketch the graph $y = \sin x$ for $0° \leqslant x \leqslant 360°$ and the line $y = \dfrac{1}{3}$. Indicate the solution of this equation that you found on your calculator (see Figure 2).

Handy hint

You need to re-express the equation in the form $\sin x = k$ where k is a constant.

A Level Alert!

You need to find one solution of the given equation using your calculator.

Common error

The answer is **not** $\sin^{-1}(1) \div 3$.

Brackets in the correct place are essential here!

Figure 2

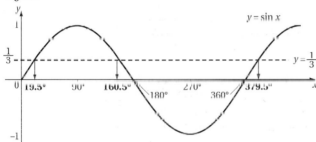

You can see that other values of x where this line and curve intersect are:

$\quad x = 180° - 19.5° \qquad x = 360° + 19.5°$

$\quad = 160.5° \qquad\qquad = 379.5°$

The value $x = 379.5°$ does not lie in the range $0° \leqslant x \leqslant 360°$ so it must be ignored.

Hence, to 1 decimal place, $x = 19.5°$ and $x = 160.5°$ are the solutions of the equation $3 \sin x + 2 = 3$ for $0° \leqslant x \leqslant 360°$.

Handy hint

Take care to include only those values in the given range, in this case $0° \leqslant x \leqslant 360°$.

Taking it further

At A Level, you will need to solve trigonometric equations such as $2 \sin (2x + 30°) = 1$ for $0° \leqslant x \leqslant 360°$. The techniques in this section can be used to solve equations like this.

7 Vectors

7.1 The magnitude and direction of a vector

What you should already know:

*   how to add and subtract column vectors, for example: $\begin{pmatrix} 3 \\ 2 \end{pmatrix} + \begin{pmatrix} 4 \\ -1 \end{pmatrix} - \begin{pmatrix} 2 \\ 1 \end{pmatrix}$.

*   how to multiply a column vector by a number, for example: $3\begin{pmatrix} 2 \\ -5 \end{pmatrix}$.

In this section you will learn:

*   how to find the magnitude and direction of a vector.
*   how to write a vector in $\mathbf{i} - \mathbf{j}$ form.

⟫ **Revisiting GCSE** ⟫⟫

You know that the column vector $\begin{pmatrix} 2 \\ 3 \end{pmatrix}$ means 'move 2 steps right and 3 steps up' and that you can use bold lower-case letters to represent vectors.

GCSE Example 1 _____

The vectors $\mathbf{a} = \begin{pmatrix} 4 \\ 3 \end{pmatrix}$ and $\mathbf{b} = \begin{pmatrix} 3 \\ -4 \end{pmatrix}$
Work out:

a $2\mathbf{a} + \mathbf{b}$

b $3\mathbf{a} - 2\mathbf{b}$.

Working

a You work out $2\mathbf{a}$ by multiplying each number in vector $\mathbf{a}$ by 2.

$$\mathbf{a} = \begin{pmatrix} 4 \\ 3 \end{pmatrix} \qquad \text{so} \qquad 2\mathbf{a} = \begin{pmatrix} 2 \times 4 \\ 2 \times 3 \end{pmatrix} = \begin{pmatrix} 8 \\ 6 \end{pmatrix}$$

So $2\mathbf{a} + \mathbf{b} = \begin{pmatrix} 8 \\ 6 \end{pmatrix} + \begin{pmatrix} 3 \\ -4 \end{pmatrix}$

$= \begin{pmatrix} 8+3 \\ 6-4 \end{pmatrix}$

$= \begin{pmatrix} 11 \\ 2 \end{pmatrix}$

b $3\mathbf{a} - 2\mathbf{b} = 3\begin{pmatrix} 4 \\ 3 \end{pmatrix} - 2\begin{pmatrix} 3 \\ -4 \end{pmatrix}$

$= \begin{pmatrix} 12 \\ 9 \end{pmatrix} - \begin{pmatrix} 6 \\ -8 \end{pmatrix}$

$= \begin{pmatrix} 12-6 \\ 9-(-8) \end{pmatrix}$

$= \begin{pmatrix} 6 \\ 17 \end{pmatrix}$

You can draw a directed line to represent the vector $\mathbf{a} = \begin{pmatrix} 2 \\ 3 \end{pmatrix}$.

Since $\begin{pmatrix} 2 \\ 3 \end{pmatrix}$ means 'move 2 steps right and 3 steps up', this vector has **horizontal component** 2 and **vertical component** 3.

A vector is completely described by its **magnitude** and **direction**. The diagram shows the vector **a**.

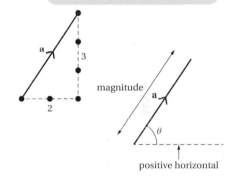

Key point

- The **magnitude** of **a** is the length of the line representing **a**.
- You write $|\mathbf{a}|$ for the magnitude of **a**.
- The **direction** of **a** is the angle θ between this line and the positive horizontal.
- You can assume all directions are measured against the positive horizontal and lie in the range $0° \le \theta, 360°$.

You use the components of a vector to calculate its magnitude and direction.

A Level Example 2

The vector $\overrightarrow{AB} = \begin{pmatrix} 3 \\ 2 \end{pmatrix}$.

a   Find the magnitude of $\overrightarrow{AB}$, giving the answer to 1 decimal place.

b   Find the direction of $\overrightarrow{AB}$, giving the answer to the nearest degree.

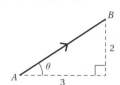

Working

Sketch the vector $\overrightarrow{AB} = \begin{pmatrix} 3 \\ 2 \end{pmatrix}$, showing its components and direction θ.

The triangle shown is right-angled.

a Use Pythagoras' theorem to calculate the magnitude of $\overrightarrow{AB}$.

$$|\overrightarrow{AB}| = \sqrt{3^2 + 2^2}$$
$$= \sqrt{13}$$
$$= 3.605\ldots$$
$$|\overrightarrow{AB}| = 3.6 \text{ (1 decimal place)}.$$

b Use trigonometry to calculate angle θ.

$$\tan\theta = \frac{\text{opposite}}{\text{adjacent}}$$
$$= \frac{2}{3}$$

So $\theta = \tan^{-1}\left(\dfrac{2}{3}\right)$
$$= 33.69\ldots°$$

$\overrightarrow{AB}$ has direction 34° (nearest degree).

Common error

$\sqrt{3^2 + 2^2}$ is **not** equal to $3 + 2$.

Add the squares then square root the answer.

Handy hint

Use tan because you are given the opposite and adjacent sides to θ in the triangle.

> **Key point**

The vector $\begin{pmatrix} x \\ y \end{pmatrix}$ has:

- magnitude $\left|\begin{pmatrix} x \\ y \end{pmatrix}\right| = \sqrt{x^2 + y^2}$

- direction θ, where $\tan\theta = \dfrac{y}{x}$, $x \neq 0$.

At A Level, vectors are also written using a combination of the vectors $\mathbf{i} = \begin{pmatrix} 1 \\ 0 \end{pmatrix}$ and $\mathbf{j} = \begin{pmatrix} 0 \\ 1 \end{pmatrix}$.

For example, you can write $\begin{pmatrix} 3 \\ 4 \end{pmatrix}$ as $3 \times \begin{pmatrix} 1 \\ 0 \end{pmatrix} + 4 \times \begin{pmatrix} 0 \\ 1 \end{pmatrix}$, or $3\mathbf{i} + 4\mathbf{j}$.

$3\mathbf{i} + 4\mathbf{j}$ is the vector $\begin{pmatrix} 3 \\ 4 \end{pmatrix}$ written in $\mathbf{i} - \mathbf{j}$ form.

> **Key point**

$$\begin{pmatrix} x \\ y \end{pmatrix} = x\mathbf{i} + y\mathbf{j}$$

column vector $\mathbf{i} - \mathbf{j}$ form

The vector $x\mathbf{i} + y\mathbf{j}$ has:

- magnitude $|x\mathbf{i} + y\mathbf{j}| = \sqrt{x^2 + y^2}$

- direction θ, where $\tan\theta = \dfrac{y}{x}$, $x \neq 0$.

A Level Example 3

For the vectors $\mathbf{a} = 2\mathbf{i} - 3\mathbf{j}$, $\mathbf{b} = 4\mathbf{i} + 8\mathbf{j}$ and $\mathbf{c} = 4\mathbf{i} + p\mathbf{j}$, it is given that $3\mathbf{a} + 2\mathbf{b} - q\mathbf{c} = \mathbf{0}$, where p and q are constants.
Find the magnitude of $\mathbf{c}$, giving the answer in simplified surd form.

Handy hint

$\mathbf{0} = 0\mathbf{i} + 0\mathbf{j}$ is the zero vector.

Working

Find the components of $3\mathbf{a} + 2\mathbf{b} - q\mathbf{c}$.

$$3\mathbf{a} + 2\mathbf{b} - q\mathbf{c} = 3(2\mathbf{i} - 3\mathbf{j}) + 2(4\mathbf{i} + 8\mathbf{j}) - q(4\mathbf{i} + p\mathbf{j})$$
$$= (6\mathbf{i} - 9\mathbf{j}) + (8\mathbf{i} + 16\mathbf{j}) - (4q\mathbf{i} + qp\mathbf{j})$$

Combine like components: $= (6 + 8 - 4q)\mathbf{i} + (-9 + 16 - qp)\mathbf{j}$

Simplify: $= (14 - 4q)\mathbf{i} + (7 - qp)\mathbf{j}$

So $(14 - 4q)\mathbf{i} + (7 - qp)\mathbf{j} = 0\mathbf{i} + 0\mathbf{j}$

Equate $\mathbf{i}$ components: $14 - 4q = 0$

Solve for q: $\qquad\qquad q = 3.5$

Equate $\mathbf{j}$ components: $7 - 3.5p = 0$

Solve for p: $\qquad p = 2$

So $\mathbf{c} = 4\mathbf{i} + 2\mathbf{j}$ and hence $|\mathbf{c}| = \sqrt{4^2 + 2^2}$
$$= \sqrt{20}$$
$$= 2\sqrt{5}$$

Handy hint

You may find it helpful to write $\mathbf{a}$, $\mathbf{b}$ and $\mathbf{c}$ as column vectors.

Handy hint

Use the value of q to find the value of p.

At A Level, particularly when studying Mechanics, it is important to be able to find the components of a vector from its magnitude and direction.

A Level Example 4

The vector $\mathbf{q} = -4\mathbf{i} + \mathbf{j}$ and the vector $\mathbf{r}$ has magnitude 4 and direction $65°$.
Find the magnitude and direction of $\mathbf{q} + \mathbf{r}$. Give answers to 1 decimal place.

Working

You first need to find the components of $\mathbf{r}$.

Let $\mathbf{r} = x\mathbf{i} + y\mathbf{j}$ where components x and y are to be found.

The sketch shows the magnitude and direction of $\mathbf{r}$.

Use trigonometry on this right-angled triangle to find x and y.

$$\cos 65° = \frac{x}{4} \text{ so } x = 4 \cos 65°$$
$$= 1.6904\ldots$$
$$\sin 65° = \frac{y}{4} \text{ so } y = 4 \sin 65°$$
$$= 3.6252\ldots$$

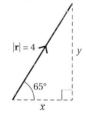

So, to 2 decimal places, $\mathbf{r} = 1.69\mathbf{i} + 3.63\mathbf{j}$.

Working to 2 decimal places, calculate the components of $\mathbf{q} + \mathbf{r}$.

$$\mathbf{q} + \mathbf{r} = (-4\mathbf{i} + \mathbf{j}) + (1.69\mathbf{i} + 3.63\mathbf{j})$$
$$= -2.31\mathbf{i} + 4.63\mathbf{j}$$

So $|\mathbf{q} + \mathbf{r}| = \sqrt{(-2.31)^2 + 4.63^2}$
$$= \sqrt{26.773}$$
$$= 5.1742\ldots$$
$$|\mathbf{q} + \mathbf{r}| = 5.2 \ (1 \text{ decimal place})$$

Sketch the vector $\mathbf{q} + \mathbf{r}$ to find its direction θ.

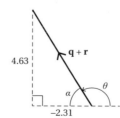

Handy hint

Round components to 2 decimal places so that the final answers are correct to 1 decimal place.

Common error

Do **not** enter $\sqrt{-2.31^2 + 4.63^2}$ on your calculator. Use brackets with negative numbers.

Common error

$|\mathbf{q} + \mathbf{r}|$ is **not** equal to $|\mathbf{q}| + |\mathbf{r}|$ because $\mathbf{q}$ and $\mathbf{r}$ have different directions.

From the sketch, $\theta = 180° - \alpha$ where the acute angle α is such that
$$\tan \alpha = \frac{4.63}{2.31}$$

Calculate α:
$$\alpha = \tan^{-1}\left(\frac{4.63}{2.31}\right)$$
$$= 63.4844...°$$

Calculate θ:
$$\theta = 180° - \alpha$$
$$= 180° - 63.4844...°$$
$$= 116.5155...°$$

$\mathbf{q} + \mathbf{r}$ has direction $116.5°$ (1 decimal place).

> **Common error**
>
> $\tan \alpha$ is **not** equal to $\frac{4.63}{-2.31}$
> To find α, ignore signs on the components.

Key point

- The vector $\mathbf{a}$ with direction θ has:
  - horizontal component $|\mathbf{a}|\cos \theta$
  - vertical component $|\mathbf{a}|\sin \theta$.
- Always sketch a vector when finding its direction.

The diagrams explain how to find the direction θ of a vector $\mathbf{a}$ when θ is not acute. In each diagram, α is an acute angle.

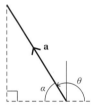

$90° < \theta < 180°$

Direction: $\theta = 180° - \alpha$

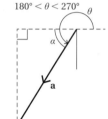

$180° < \theta < 270°$

Direction: $\theta = 180° + \alpha$

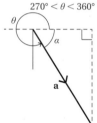

$270° < \theta < 360°$

Direction: $\theta = 360° - \alpha$

Taking it further

You will use vectors to solve problems in Mechanics. Forces are vector quantities and can be added or subtracted by writing them in component form.

7.2 Position vectors

What you should already know:
- how to use the triangle law of vector addition: $\overrightarrow{AB} + \overrightarrow{BC} = \overrightarrow{AC}$.
- how to solve geometrical problems using vectors.

In this section you will learn:
- how to find the position vector of a point.
- how to use position vectors to find the distance between two points and solve geometrical problems.

At GCSE, you solved geometrical problems using vectors.

GCSE Example 5

In triangle ABC, M is the midpoint of AB and N is the midpoint of AC.

a Show that MN is parallel to BC.

b Find the ratio of the area of triangle AMN to the area of the trapezium $MNCB$.

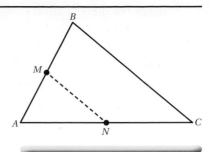

Working

a You need to express $\overrightarrow{MN}$ in terms of the vector $\overrightarrow{BC}$.

M is the midpoint of AB so $\overrightarrow{MA} = \frac{1}{2}\overrightarrow{BA}$.

Similarly, $\overrightarrow{AN} = \frac{1}{2}\overrightarrow{AC}$.

By the triangle law of addition:

$$\overrightarrow{MN} = \overrightarrow{MA} + \overrightarrow{AN}$$
$$= \frac{1}{2}\overrightarrow{BA} + \frac{1}{2}\overrightarrow{AC}$$
$$= \frac{1}{2}\left(\overrightarrow{BA} + \overrightarrow{AC}\right)$$

Now use the triangle law again: $= \frac{1}{2}\overrightarrow{BC}$

> **Handy hint**
>
> Use vectors to express the information you are given about M and N.

> **Handy hint**
>
> Think of + as meaning 'followed by'
> $\overrightarrow{MA}$ **followed by** $\overrightarrow{AN}$ is equivalent to $\overrightarrow{MN}$.

So $\overrightarrow{MN} = \frac{1}{2}\overrightarrow{BC}$. This means MN has the same direction as BC but is half its length. This proves MN is parallel to BC.

b From the result in part **a**, $\triangle ABC$ is a scale factor 2 enlargement of $\triangle AMN$.

So area $\triangle ABC = 4 \times$ area $\triangle AMN$.

Hence, area of $MNCB = 3 \times$ area $\triangle AMN$,

Triangle AMN and trapezium $MNCB$ have areas in the ratio $1:3$.

> **Handy hint**
>
> Area scale factor = (length scale factor)².

At A Level, you need to work with **position** vectors.

> **Key point**
>
> - The position vector of the point P is the vector $\overrightarrow{OP}$, where O is the origin.
> - $\overrightarrow{OP} = \begin{pmatrix} a \\ b \end{pmatrix}$ is equivalent to saying P has coordinates (a, b).

For example, the point with coordinates $P(2, 3)$ has position vector $\overrightarrow{OP} = \begin{pmatrix} 2 \\ 3 \end{pmatrix}$, as shown in the diagram.

Using position vectors, you can find the vector joining two points.

If A and B are any two points then, by the triangle law of addition, $\overrightarrow{OA} + \overrightarrow{AB} = \overrightarrow{OB}$.

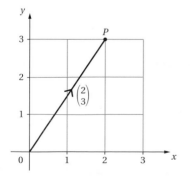

You can rewrite this equation as: $\overrightarrow{AB} = \overrightarrow{OB} - \overrightarrow{OA}$.

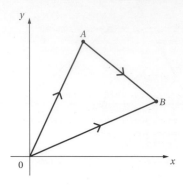

> ## Key point
>
> For any two points A and B, $\overrightarrow{AB} = \overrightarrow{OB} - \overrightarrow{OA}$.

You can use this result to solve geometrical problems.

A Level Example 6

Relative to the origin O, points A and B have coordinates $(1, 2)$ and $(4, -2)$, respectively.

a Find $\left|\overrightarrow{AB}\right|$.

Point C has coordinates $(13, k)$, where k is a positive constant. Given that $\left|\overrightarrow{AC}\right| = 13$:

b find the exact perimeter of triangle ABC.

A Level Alert!

The word 'respectively' is used frequently at A Level. It means you pair items off in the obvious order, so here, $A(1, 2)$ and $B(4, -2)$.

Working

a First find the components of $\overrightarrow{AB}$.

$\overrightarrow{OA} = \begin{pmatrix} 1 \\ 2 \end{pmatrix}$ and $\overrightarrow{OB} = \begin{pmatrix} 4 \\ -2 \end{pmatrix}$ are the position vectors of A and B, respectively.

$$\overrightarrow{AB} = \overrightarrow{OB} - \overrightarrow{OA}$$

$$= \begin{pmatrix} 4 \\ -2 \end{pmatrix} - \begin{pmatrix} 1 \\ 2 \end{pmatrix}$$

$$= \begin{pmatrix} 3 \\ -4 \end{pmatrix}$$

Checkpoint

Verify this answer by finding the length AB using the distance formula – see Section 3.3.

So: $\left|\overrightarrow{AB}\right| = \sqrt{3^2 + (-4)^2}$

$$= \sqrt{25}$$

$$= 5$$

b Sketch triangle ABC, showing all known lengths.
Use the information $\left|\overrightarrow{AC}\right| = 13$ to find k:

$$\overrightarrow{AC} = \overrightarrow{OC} - \overrightarrow{OA}$$

$$= \begin{pmatrix} 13 \\ k \end{pmatrix} - \begin{pmatrix} 1 \\ 2 \end{pmatrix}$$

$$= \begin{pmatrix} 12 \\ k-2 \end{pmatrix}$$

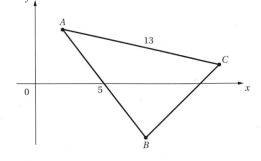

Express $|\overline{AC}| = 13$ as an equation involving k:

$$|\overline{AC}| = \sqrt{12^2 + (k-2)^2}$$

So: $\qquad \sqrt{12^2 + (k-2)^2} = 13$

Square each side: $12^2 + (k-2)^2 = 13^2$

So: $\qquad\qquad (k-2)^2 = 169 - 144$

$$= 25$$

Square root each side: $\qquad k - 2 = \pm 5$

Solve for k: $\qquad\qquad k = 7$ or $k = -3$

So $k = 7$, because k is positive.

Hence, $\overline{AC} = \begin{pmatrix} 12 \\ 7-2 \end{pmatrix}$

$$= \begin{pmatrix} 12 \\ 5 \end{pmatrix}$$

Handy hint

You can still sketch the triangle even though you do not know where to place C.

Handy hint

You could solve the equation $(k-2)^2 = 25$ by expanding the brackets to form a quadratic, but taking square roots is quicker.

Handy hint

Underline and make use of all the important information in the question (in this case, that k is positive).

Now find $\overline{BC}$ so that you can calculate $|\overline{BC}|$.

$$\overline{BC} = \overline{BA} + \overline{AC}$$

$$= -\overline{AB} + \overline{AC} \quad \text{where} \quad \overline{AB} = \begin{pmatrix} 3 \\ -4 \end{pmatrix}$$

$$= \begin{pmatrix} -3 \\ 4 \end{pmatrix} + \begin{pmatrix} 12 \\ 5 \end{pmatrix}$$

$$= \begin{pmatrix} 9 \\ 9 \end{pmatrix}$$

So: $\quad |\overline{BC}| = \sqrt{9^2 + 9^2}$

$$= 9\sqrt{2}$$

The perimeter of triangle $ABC = 5 + 9\sqrt{2} + 13$

$$= 18 + 9\sqrt{2} \text{ units.}$$

You can use position vectors to prove geometric formulae.

Handy hint

$\overline{BA} = -\overline{AB}$ since BA and AB have equal lengths but opposite directions.

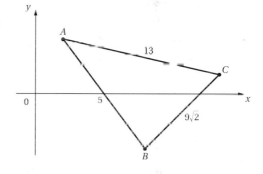

A Level Example 7

The diagram shows points A and B with coordinates (x_1, y_1) and (x_2, y_2), respectively.

M is the mid-point of AB.

Use vectors to prove that the coordinates of M are $\left(\dfrac{x_1 + x_2}{2}, \dfrac{y_1 + y_2}{2} \right)$.

Working

$$\overline{OA} = \begin{pmatrix} x_1 \\ y_1 \end{pmatrix} \text{ and } \overline{OB} = \begin{pmatrix} x_2 \\ y_2 \end{pmatrix}.$$

M is the mid-point of AB so $\overline{AM} = \frac{1}{2}\overline{AB}$.

You need to find an expression for the position vector $\overline{OM}$.

By the triangle law: $\qquad\qquad \overline{OM} = \overline{OA} + \overline{AM}$

$$= \overline{OA} + \frac{1}{2}\overline{AB}$$

Use $\overline{AB} = \overline{OB} - \overline{OA}$: $\qquad = \overline{OA} + \frac{1}{2}\left(\overline{OB} - \overline{OA}\right)$

$$= \frac{1}{2}\overline{OA} + \frac{1}{2}\overline{OB}$$

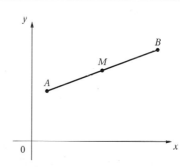

Handy hint

Where possible, work with vectors rather than their components.

Simplify:
$$= \frac{1}{2}\left(\overline{OA} + \overline{OB}\right)$$

Use components of $\overline{OA}$ and $\overline{OB}$:
$$= \frac{1}{2}\left(\begin{pmatrix} x_1 \\ y_1 \end{pmatrix} + \begin{pmatrix} x_2 \\ y_2 \end{pmatrix}\right)$$

So $\overline{OM} = \frac{1}{2}\begin{pmatrix} x_1 + x_2 \\ y_1 + y_2 \end{pmatrix}$ which proves that the coordinates of M are $\left(\dfrac{x_1 + x_2}{2}, \dfrac{y_1 + y_2}{2}\right)$.

You might like to compare the proof about mid-points in this example with that given in Section 3.3.

Taking it further

As you progress through your course you will study vectors in three dimensions. You will also learn how to use vectors to solve problems involving projectile motion.

8 Differentiation

8.1 Estimating the gradient of a curve

What you should already know:
- how to find the gradient of a line passing through two points $P(x_1, y_1)$ and $Q(x_2, y_2)$.
- how to estimate the gradient of a curve by drawing a tangent

In this section you will learn:
- how to find the **exact** gradient of a curve at any point.

> **Revisiting GCSE**

What you should already know:
- how to find the gradient of a line given the coordinates of two points on the line.
- how to estimate the gradient of a curve at a point by drawing a tangent.
- how to interpret the gradient of a curve as a rate of change.

In this section you will learn:
- how to estimate the gradient of a curve at a point by using a chord of the curve.

You already know how to estimate the gradient of a curve at a point P, by drawing a tangent to the curve at P and calculating its slope.

You may have applied this technique to a distance–time graph or a speed–time graph and then been asked to interpret what the gradient means.

The diagram shows the distance, in metres, travelled by a car in t seconds after passing a road marker used by a speed camera.

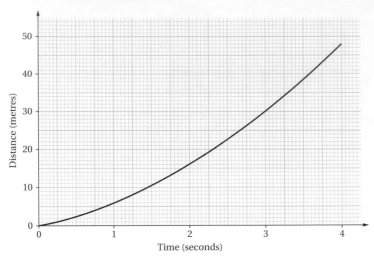

a Calculate an estimate for the gradient of this curve when $t = 2$.

b What information about the car does this estimate provide?

Working

a Carefully draw the tangent to the curve at the point P where $t = 2$.

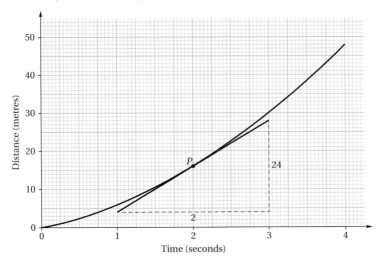

This tangent passes through the points (1, 4) and (3, 28).

Draw a right-angled triangle as shown.

This triangle has base 2 and height 24.

So the gradient of this tangent $= \dfrac{\text{height}}{\text{base}} = \dfrac{24}{2}$
$$= 12$$

When $t = 2$, the gradient of the curve is approximately 12.

b The gradient of a distance–time graph represents speed.
The speed of the car 2 seconds after passing the marker was approximately 12 metres per second.

> **Handy hint**
> The answer is an estimate because the tangent was drawn 'by eye'.

> **Handy hint**
> A gradient measures a rate of change. Speed is the rate at which distance is changing.

Moving on to A Level

At A Level, finding an approximate gradient is not very useful. You need to be able to find the **exact** gradient of a curve at any point.

Since it is difficult to draw the tangent to a curve by hand, you need a theoretical method for finding a gradient which does not rely on measurement or judgement.

The following examples will help you understand this theoretical method, which is explained in section 8.2.

Figure 1 shows the curve with equation $y = x^2$ passing through the points $P(1, 1)$ and $Q(2, 4)$.

The tangent, T, to this curve at P has also been drawn.

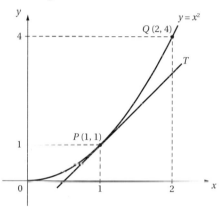

Figure 1

To find the gradient of a line, you need to know the coordinates of *two* points on the line.

Since the diagram has not been drawn accurately, you do not have enough information to find the gradient of T.

However, by using the points $P(1, 1)$ and $Q(2, 4)$, you *can* find the gradient of the line PQ, which is a reasonable estimate for the gradient of T (see Figure 2).

$$\text{Grad}_{PQ} = \frac{y_2 - y_1}{x_2 - x_1}$$

$$= \frac{4 - 1}{2 - 1}$$

$$= 3$$

So the gradient of T is approximately 3.

Since P and Q lie on the curve, the line segment PQ is called a **chord** of the curve.

To obtain a better estimate for the gradient of T you can use, for example, the chord PQ', where the point $Q'(1.8, 3.24)$ on this curve is *closer* than Q to point P (see Figure 3).

$$\text{Grad}_{PQ'} = \frac{y_2 - y_1}{x_2 - x_1}$$

$$= \frac{3.24 - 1}{1.8 - 1}$$

$$= 2.8$$

So a better estimate for the gradient of T is 2.8.

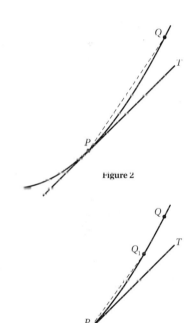

Figure 2

Figure 3

Handy hint

Point Q' lies on the curve $y = x2$ so its y–coordinate is $1.8^2 = 3.24$.

Handy hint

Grad_T means the gradient of the tangent T.

The symbol $\approx$ means 'is approximately equal to'.

DIFFERENTIATION **61**

> ## Key point
>
> Let P be a fixed point on a curve and T the tangent to this curve at P.
> Gradient of curve at $P = \text{Grad}_T \approx \text{Grad}_{PQ}$ where Q is any other point on this curve.
> The closer Q is to P, the better this approximation.

For example, if instead you used the point $Q''(1.1, 1.21)$ on this curve, then $\frac{1.21-1}{1.1-1} = 2.1$ is a very good estimate for the gradient of the curve with equation $y = x^2$ at the point $P(1, 1)$.

A Level Example 2

The curve with equation $y = x^2$ passes through the point P with x-coordinate 3 and point Q with x-coordinate $(3 + h)$, where h is a positive constant.

Use the chord PQ to find an expression, in terms of h, which approximates the gradient of this curve at point P.

Working

Sketch a diagram showing the coordinates of points P and Q, and the tangent T at P.

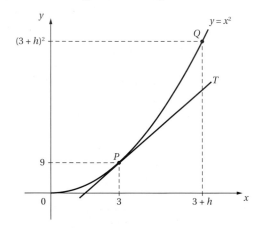

The equation of the curve is $y = x^2$ so,
the y-coordinate of point P is $3^2 = 9$
and,
the y-coordinate of point Q is $(3 + h)^2$.

> *Handy hint*
> The gradient of the curve at a point equals the gradient of the tangent to the curve at that point.

Using the chord PQ: gradient of curve at $P \approx \text{Grad}_{PQ}$

$$= \frac{y_2 - y_1}{x_2 - x_1}$$

$$= \frac{(3+h)^2 - 9}{(3+h) - 3}$$

Expand $(3 + h)^2$ and simplify the denominator: $= \dfrac{9 + 6h + h^2 - 9}{h}$

Simplify and factorise the numerator: $= \dfrac{h(6 + h)}{h}$

Cancel the h terms: $= 6 + h$

So the gradient of the curve with equation $y = x^2$ at the point $P(3, 9)$ is approximately $6 + h$.

Where this topic goes next:
You will learn how the *exact* gradient of a curve at any point can be found.

8.2 The rules of differentiation

What you should already know:

- how to estimate the gradient of curve at a point P by calculating Grad_{PQ}, where Q is a point on the curve which is close to P.

In this section you will learn:

- how to find the exact gradient of a curve at any point by using differentiation from first principles.

You can use the method of Example 2 to find the gradient of a curve at *any* point.

A Level Example 3

Show that the gradient of the curve with equation $y = x^2$ at the point $P(x, x^2)$ is $2x$.

Working

Points $P(x, x^2)$ and $Q(x + h, (x + h)^2)$, where $h > 0$, lie on this curve.

The gradient of the chord PQ is a good estimate for the gradient of the curve at P.

Using the chord PQ: gradient of curve at $P \approx \text{Grad}_{PQ}$

$$= \frac{(x + h)^2 - x^2}{(x + h) - x}$$

Expand $(x + h)^2$, simplify the denominator: $= \dfrac{x^2 + 2xh + h^2 - x^2}{h}$

Simplify and factorise the numerator: $= \dfrac{h(2x + h)}{h}$

Cancel the h terms: $= 2x + h$

The gradient of this curve at P is approximately $2x + h$.

As h approaches 0:

1 Q moves closer to P and so $2x + h$ approaches the gradient of the curve at point P, and

2 the expression $2x + h$ approaches the fixed quantity $2x$.

Combining results **1** and **2** shows that the gradient of the curve with equation $y = x^2$ at point $P(x, x^2)$ is $2x$.

So, for example, the gradient of this curve at the point where $x = 1$ is $2(1) = 2$.

You can use the notation $\dfrac{dy}{dx}$ for the gradient of a curve at any point.

Example 3 shows that for the curve $y = x^2$, $\dfrac{dy}{dx} = 2x$.

> **Handy hint**
> $P(x, x^2)$ is a general point on this curve, fixed once chosen.

> **Handy hint**
> You have to use the variable point Q on this curve. Its position depends on h.

> **Handy hint**
> 'As h approaches 0' means h gets smaller and smaller, for example, $h = 0.1, 0.01, 0.001, \ldots$ You write $h \to 0$.

> **Checkpoint**
> Compare this answer with the estimates found for this gradient in section 8.1.

> **Handy hint**
> $\dfrac{dy}{dx}$ is pronounced 'dy by dx'
> d means 'change' so $\dfrac{dy}{dx}$ means 'change in y over change in x', that is, a gradient.

> **Key point**
>
> This process of finding an expression for $\dfrac{dy}{dx}$ is called **differentiation from first principles**.
>
> $\dfrac{dy}{dx}$ is **not** a fraction, so you cannot cancel the 'd's!

A Level Example 4

a Expand and then simplify $(x + h)^3$.

b Hence show, from first principles, that if $y = x^3$ then $\dfrac{dy}{dx} = 3x^2$.

Handy hint

'Hence' means 'Making use of the result from part **a**'.

Working

a Start by expanding $(x + h)^2$: $(x + h)^2 = x^2 + 2xh + h^2$

So:
$$(x + h)^3 = (x + \boldsymbol{h})(x + h)^2$$
$$= (\boldsymbol{x} + \boldsymbol{h})(x^2 + 2xh + h^2)$$
$$= \boldsymbol{x}(x^2 + 2xh + h^2) + \boldsymbol{h}(x^2 + 2xh + h^2)$$

Expand each bracket:
$$= x^3 + 2x^2h + xh^2 + hx^2 + 2xh^2 + h^3$$

Combine like terms:
$$= x^3 + 3x^2h + 3xh^2 + h^3$$

So $(x + h)^3 = x^3 + 3x^2h + 3xh^2 + h^3$.

b Points $P(x, x^3)$ and $Q(x + h < (x + h)^3)$, where $h > 0$, lie on this curve.

$$\frac{dy}{dx} \approx \text{Grad}_{PQ}$$
$$= \frac{(x+h)^3 - x^3}{(x+h) - x}$$

Use the result of part **a**:
$$= \frac{(x^3 + 3x^2h + 3xh^2 + h^3) - x^3}{h}$$

Simplify and factorise the numerator:
$$= \frac{h(3x^2 + 3xh + h^2)}{h}$$

Cancel the h terms:
$$= 3x^2 + 3xh + h^2$$

As $h \to 0$, $(3x^2 + 3xh + h^2) \to 3x^2$

So $\dfrac{dy}{dx} = 3x^2$.

You should try Question 1 and Question 2 in Practice 8.2 before moving on with this section.

Differentiation from first principles can be a lengthy process. Fortunately, there is a rule which can be used to differentiate x^n where n is any real number.

The table shows some equations and their **derivatives** (that is, expressions for $\dfrac{dy}{dx}$) with respect to x found by using first principles.

Handy hint

When you *differentiate* $y = x^4$ the answer is $\dfrac{dy}{dx} = 4x^3$.

You say the *derivative* of x^4 (with respect to x) is $4x^3$.

Curve equation	Derivative
$y = x^2$	$\dfrac{dy}{dx} = 2x$ (see Example 3)
$y = x^3$	$\dfrac{dy}{dx} = 3x^2$ (see Example 4)
$y = x^4$	$\dfrac{dy}{dx} = 4x^3$ (see Practice 8.2 Q2)

These results suggest the following rule for differentiating x^n.

⟫ Key point ⟫

The $(n - 1)$ rule: if $y = x^n$, where n is a constant, then $\dfrac{dy}{dx} = nx^{n-1}$.

Without using the letter y, you can express this rule as: the derivative of x^n with respect to x is nx^{n-1}.

Checkpoint

Verify this rule works for the equations in the table.

Check that it also works when $n = 1$ and when $n = 0$.

You can use the $(n - 1)$ rule to quickly differentiate more complicated equations.

a Differentiate $2x^5 + \frac{2}{3}x^3 - 3x + 5$ with respect to x.

b Find the gradient of the curve C with equation $y = (x^2 + 4)(2x - 3)$ at the point P where this curve crosses the x-axis.

Working

a Write each term in the expression in the form $k(x^n)$ for constants k and n.

Differentiate the bracketed terms using the $(n - 1)$ rule:

$$2(x^5) + \frac{2}{3}(x^3) - 3(x^1) + 5(x^0)$$

$$\downarrow \qquad \downarrow \qquad \downarrow \qquad \downarrow$$

$$2(5x^4) + \frac{2}{3}(3x^2) - 3(1x^0) + 5(0x^{-1})$$

Simplify: $-10x^4 + 2x^2 - 3$

So the derivative with respect to x of $2x^5 + \frac{2}{3}x^3 - 3x + 5$ is $10x^4 + 2x^2 - 3$.

b You need to differentiate the equation of the curve.

Curve equation: $\qquad y = (x^2 + 4)(2x - 3)$

Expand the brackets: $\qquad = 2x^3 - 3x^2 + 8x - 12$

Use the $(n - 1)$ rule: $\dfrac{dy}{dx} = 6x^2 - 6x + 8$

You need to find the x-coordinate of the point P

When $y = 0$, $(x^2 + 4)(2x - 3) = 0$.

So $x^2 + 4 = 0$ or $2x - 3 = 0$.

Solve for x: $\qquad x = \sqrt{-4}$ or $x = \frac{3}{2}$

So the x-coordinate of P is $\frac{3}{2}$.

Evaluate $\dfrac{dy}{dx}$ when $x = \frac{3}{2}$: $\dfrac{dy}{dx} = 6x^2 - 6x + 8$

$$= 6\left(\frac{3}{2}\right)^2 - 6\left(\frac{3}{2}\right) + 8$$

$$= 12.5$$

The gradient of C at its x-intercept is 12.5.

Here is a sketch of this curve.

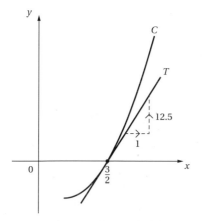

The gradient of the tangent to C at its x-intercept is 12.5.

Handy hint

$x^1 = x$ and $x^0 = 1$.

Handy hint

You bring down ($\downarrow$) the coefficients and constant term.

Handy hint

$3(1x^0) = 3 \times 1 = 3$

$5(0x^{-1}) - 5 \times \dfrac{0}{x} - 0$

Handy hint

$2(x^3) \xrightarrow{\text{differentiates to}} 2(3x^2) = 6x^2$

Handy hint

Use the factorised form of the curve equation to find its roots – see Section 5.2.

Handy hint

$\sqrt{-4}$ is not a real number so ignore this answer.

Handy hint

'Evaluate' means 'work out the value of'.

> **Taking it further** »»»»

You will learn how to use differentiation to find the equation of a tangent to a curve. Differentiation also has uses in practical problems such as maximising volumes of containers.

8.3 Applying differentiation

What you should already know:

- how to use the nx^{n-1} rule to differentiate powers of x.
- how to calculate the gradient of a curve at any given point.
- how to find an equation for a line in the form $y - y_1 = m(x - x_1)$.

In this section you will learn:

- how to find the equation of the tangent and the normal to a curve at any point.
- how to locate stationary points on a curve.

Any tangent to a curve at a point $P(x_1, y_1)$ is a straight line which passes through P.

Handy hint
See Section 3.2.

So, an equation for this tangent is $y - y_1 = m(x - x_1)$, where m is its gradient.

You can use differentiation to find the value of m.

A Level Example 6

The diagram shows the curve with equation $y = 1 + 4x - x^2$ which passes through the point $P(3, 4)$.

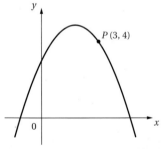

Find an equation for the tangent to this curve at point P.

Give your answer in the form $ay + bx = c$ for integers a, b and c.

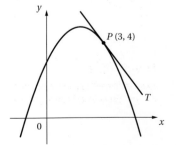

Working

The diagram shows the tangent, T, to this curve at the point P.

An equation for T is $y - y_1 = m(x - x_1)$ where $(x_1, y_1) = (3, 4)$ and m is the gradient of T.

To find m, you need to first find $\dfrac{dy}{dx}$ for this curve.

Curve equation: $y = 1 + 4x - x^2$

Differentiate using the $(n - 1)$ rule: $\dfrac{dy}{dx} = 4 - 2x$

Evaluate $\dfrac{dy}{dx}$ at the point $P(3, 4)$: $= 4 - 2(3)$

$\qquad\qquad\qquad\qquad\qquad\quad = -2$

So T passes through the point $P(3, 4)$ and has gradient -2.

An equation for T is: $\qquad\qquad y - 4 = (-2)(x - 3)$

Expand the bracket: $\qquad\qquad y - 4 = -2x + 6$

Rearrange to the required form: $y + 2x = 10$

Handy hint

Use the x-coordinate of P to find the gradient of the tangent.

Checkpoint

From the sketch you can see that the gradient of T must be negative, so $m = -2$ is a reasonable answer.

> **Key point**

An equation for the tangent to a curve at the point $P(x_1, y_1)$ is $y - y_1 = m(x - x_1)$ where m is the value of $\dfrac{dy}{dx}$ when $x = x_1$.

> **Key point**

Given a tangent, T, to a curve at a point, P, the **normal** to this curve at P is the line which

- passes through P

and

- is perpendicular to T.

You can use differentiation to find the equation of the normal to a curve at a point.

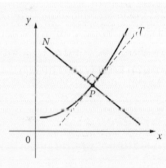

A Level Example 7

The diagram shows the curve with equation $y = \frac{1}{2}x(x^2 - 3)$.

Also shown is the normal, N, to this curve at the point P where $x = 2$.

Find the coordinates of the point A where this normally intersects the x-axis.

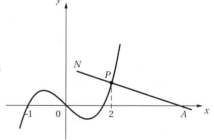

Working

To find the coordinates of point A, you first need to find the equation of N.

Use the curve equation to find the y-coordinate of point P.

When $x = 2$, $y = \frac{1}{2}(2)(2^2 - 3)$

$\qquad\qquad\qquad = 1$

Handy hint

There is no need to find the equation of T – you just need to know its gradient.

So N passes through the point $P(2, 1)$.

To find the gradient of N, you first need to find the gradient of the tangent T at point P.

Expand the bracket in the curve equation: $y = \frac{1}{2}x^3 - \frac{3}{2}x$

Differentiate the curve equation: $\quad \dfrac{dy}{dx} = \dfrac{3}{2}x^2 - \dfrac{3}{2}$

Handy hint

$\frac{1}{2}(x^3) \xrightarrow{\text{differentiates to}} \frac{1}{2}(3x^2) = \frac{3}{2}x^2$

Evaluate $\dfrac{dy}{dx}$ at point P where $x = 2$: $\text{Grad}_T = \dfrac{3}{2}(2)^2 - \dfrac{3}{2}$

$$= \dfrac{9}{2}$$

The tangent gradient at P is $\dfrac{9}{2}$, so the gradient of N is $-\dfrac{2}{9}$.

Handy hint

N is perpendicular to T.

$\text{Grad}_T = \dfrac{9}{2} \xrightarrow[\text{change sign}]{\text{invert}} -\dfrac{2}{9}$

$\qquad\qquad\qquad = \text{Grad}_N$

An equation for N is $y - 1 = -\dfrac{2}{9}(x - 2)$.

Rearrange this equation into a form that is easier to use.

Multiply each side by 9:	$9y - 9 = -2(x - 2)$
Expand the bracket:	$9y - 9 = -2x + 4$
Rearrange:	$9y + 2x = 13$
Let $y = 0$ to find the x-intercept:	$9(0) + 2x = 13$

So $x = \dfrac{13}{2}$.

Handy hint

Use $P(2, 1)$ for (x_1, y_1).

The coordinates of A are $\left(\dfrac{13}{2}, 0\right)$.

Common error

The answer is **not** $\dfrac{13}{2}$.

The question asked for coordinates.

Key point

An equation for the normal to a curve at the point $P(x_1, y_1)$ is $y - y_1 = -\dfrac{1}{m}(x - x_1)$ where m is value of $\dfrac{dy}{dx}$ when $x = x_1$.

Many questions at A Level are about locating the **stationary points** on a curve.

Key point

A **stationary point** on a curve is a point on the curve at which the gradient is zero.

The x-coordinate of a stationary point satisfies the equation $\dfrac{dy}{dx} = 0$.

At a stationary point, the tangent to the curve is horizontal.

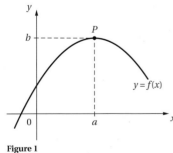

Figure 1

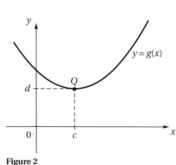

Figure 2

In Figure 1, point $P(a, b)$ is a stationary point on the curve with equation $y = f(x)$.

The value of a satisfies the equation $\dfrac{dy}{dx} = 0$.

Point P is a **maximum** point – the y-coordinates of points on the curve close to P are less than b.

Similarly, in Figure 2, the stationary point $Q(c, d)$ on the curve with equation $y = g(x)$ is a **minimum** point. The value of c is a solution to the equation $\dfrac{dy}{dx} = 0$ and the y-coordinates of points on the curve close to Q are greater than d.

A Level Example 8

The point P is the stationary point on the curve with equation $y = x^2 - 6x + 13$.

a Find the coordinates of point P.

b Hence, or otherwise, sketch this curve.

Working

a To find the x-coordinate of point P you need to solve the equation $\dfrac{dy}{dx} = 0$.

Curve equation: $y = x^2 - 6x + 13$

Differentiate: $\dfrac{dy}{dx} = 2x - 6$

Solve the equation $\dfrac{dy}{dx} = 0$: $2x - 6 = 0$

So $x = 3$.

The x-coordinate of P is 3.

> **Common error**
>
> The y-coordinate of P is **not** found by substituting $x = 3$ into the equation $\dfrac{dy}{dx} = 2x - 6$.

You now find the y-coordinate of P by substituting $x = 3$ into the curve equation.

Curve equation: $y = x^2 - 6x + 13$

Substitute in $x = 3$: $= (3)^2 - 6(3) + 13$

$= 9 - 18 + 13$

$= 4$

So the stationary point P on this curve has coordinates $(3, 4)$.

b Find the y-intercept.

When $x = 0$, $y = (0)^2 - 6(0) + 13 = 13$.

The coefficient of x^2 in the curve equation is $+1$ (that is, positive) so the graph is $\cup$-shaped, with minimum point at $P(3, 4)$.

Here is a sketch of this curve.

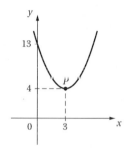

Where this topic goes next:

You will investigate stationary points in much more detail as you progress through your course. They are helpful for solving real-life problems such as maximising volumes of shapes and minimising costs of production. In particular, you will learn techniques for distinguishing between maximum and minimum points.

> **Checkpoint**
>
> Part **b** is an 'otherwise' question - you could also sketch this curve by completing the square.
>
> $y = x^2 - 6x + 13$
> $= (x - 3)^2 - 9 + 13$
> $= (x - 3)^2 + 4$
>
> This confirms the minimum point has coordinates $(3, 4)$.

9 Integration

9.1 The rules of integration

Handy hint

See Section 8.2.

What you should already know:

- how to differentiate an expression such as $x^3 - 2x^2 + 5x - 1$ with respect to x.

In this section you will learn:

- how to **integrate** an expression such as $x^3 - 2x^2 + 5x - 1$ with respect to x.
- how to evaluate an integral using its limits.

> **Key point**
>
> Integration is the process which reverses differentiation.
> The symbol for integration is $\int$.

For example, you write $\int 3x^2 \, dx$ for the integral of $3x^2$ with respect to x:

- $3x^2$ is the **integrand** (the function being integrated)
- dx means 'with respect to x'.

In order to find an answer for $\int 3x^2 \, dx$, you must think of an expression whose derivative is $3x^2$.

Clearly, x^3 is one possible answer, as $x^3 \xrightarrow{\text{differentiates to}} 3x^2 = $ integrand.

But $x^3 + 2$ is also an answer, because $(x^3 + 2) \xrightarrow{\text{differentiates to}} 3x^2 = $ integrand.

Handy hint

You met the symbol dx in Section 8.2.

Handy hint

See Section 8.2 – the derivative of any constant is zero.

The most general answer you can write down for $\int 3x^2\,dx$ is $x^3 + c$, where c represents *any* real number.

So: $\int 3x^2\,dx = x^3 + c$.

Handy hint

c is a called an *arbitrary constant*.

Key point

For any function $f(x)$, $\int f(x)\,dx = F(x) + c$, where:
- $F(x)$ is any function whose derivative with respect to x is $f(x)$.
- c is an arbitrary constant.

So, $\int x^2\,dx = \frac{1}{3}x^3 + c$ because $\frac{1}{3}x^3 \xrightarrow{\text{differentiates to}} x^2 =$ integrand

and $\int x^3\,dx = \frac{1}{4}x^4 + c$ because $\frac{1}{4}x^4 \xrightarrow{\text{differentiates to}} x^3 =$ integrand.

These two results are examples of a general rule for integrating x^n.

Key point

The $(n + 1)$ rule: $\int x^n\,dx = \frac{1}{n+1}x^{n+1} + c$ where:
- n is any number, **except** -1

and
- c is an arbitrary constant.

Checkpoint

Verify that

$\frac{1}{n+1}x^{n+1} \xrightarrow{\text{differentiates to}} x^n$.

You cannot use this rule when $n = -1$ because then, the fraction $\frac{1}{n+1}$ would equal $\frac{1}{0}$, which is undefined.

You will learn how to find $\int x^{-1}\,dx$ as you progress through your course.

Handy hint

To integrate, increase the power by 1 using the $(n + 1)$ rule.

To differentiate, decrease the power by 1 using the $(n - 1)$ rule

A Level Example 1

Find $\int (3x + 2)^2\,dx$

Handy hint

You bring down ($\downarrow$) the coefficients and constant term.

Working

Expand the brackets and simplify the answer before integrating.

Expand the brackets: $(3x + 2)^2 = 9x^2 + 12x + 4$

Express the terms in the form $k(x^n)$: $= 9(x^2) + 12(x^1) + 4(x^0)$
$\qquad\qquad\qquad\qquad\qquad\qquad\quad \downarrow \qquad\quad \downarrow \qquad\quad \downarrow$

Integrate bracketed terms using the $(n + 1)$ rule: $9\left(\frac{1}{3}x^3\right) + 12\left(\frac{1}{2}x^2\right) + 4\left(\frac{1}{1}x^1\right)$

Simplify coefficients: $= 3x^3 + 6x^2 + 4x$

So $\int (3x + 2)^2\,dx = 3x^3 + 6x^2 + 4x + c$.

Handy hint

Include '$+ c$' in your answer.

Key point

A special case of the $(n + 1)$ rule:
$\int k\,dx = kx + c$, where k is any constant.

Checkpoint

$(kx + c) \xrightarrow{\text{differentiates to}} k =$ integrand

The answer to any integration can always be checked using differentiation.

You should try Questions 1 to 3 in Practice 9.1 before moving on with this section.

You can work out the value of an integral using its **limits**.

For example, $\int_1^4 x^2 \, dx$ has **lower limit** 1 and **upper limit** 4.

An integral with limits is called a **definite** integral.

> **Handy hint**
>
> An integral without limits, such as $\int x^2 \, dx$, is called an **indefinite** integral.

> **Key point**
>
> $$\int_a^b f(x)\,dx = F(b) - F(a), \text{ where } \int f(x)\,dx = F(x) + c$$

Definite integrals are essential for working out the exact area of a region between a curve and the x-axis (see Section 9.2).

A Level Example 2

$f(x) = 2x\,(3x - 1)$

a Evaluate $\int_1^3 f(x)\,dx$.

b Find the value of the positive constant b such that $\int_0^b f(x)\,dx = 0$.

> **Handy hint**
>
> 'Evaluate' means 'find the value of'.
>
> A definite integral has a numerical value.

Working

a Start by integrating the function:
$$\int f(x)\,dx = \int 2x(3x - 1)\,dx$$

Expand the bracket: $= \int 6x^2 - 2x \, dx$

Integrate each term: $= 2x^3 - x^2 + c$
$$= F(x) + c, \text{ where } F(x) = 2x^3 - x^2$$

Now evaluate, using the limits 1 and 3:

$$\int_1^3 f(x)\,dx = F(3) - F(1)$$
$$= (2(3)^3 - (3)^2) - (2(1)^3 - (1)^2)$$
$$= (45) - (1)$$
$$= 44$$

So $\int_1^3 f(x)\,dx = 44$.

b $\int_0^b f(x)\,dx = F(b) - F(0)$

$$= (2b^3 - b^2) - (2(0)^3 - (0)^2)$$
$$= 2b^3 - b^2$$

From the information in part **b**, the numerical value of $\int_0^b f(x)\,dx$ is zero.

So: $\qquad 2b^3 - b^2 = 0$

Factorise: $\quad b^2\,(2b - 1) = 0$

> **Handy hint**
>
> If $b^2 = 0$ then $b = 0$.

Solve for b: $b = 0$ or $b = \frac{1}{2}$.

Since b is positive, $b = \frac{1}{2}$

In A Level books, the calculation $F(b) - F(a)$ is written more compactly as $\left[F(x)\right]_a^b$.

Handy hint

Use all the information in question ($b > 0$).

Limits can be fractions or negative numbers.

Key point

$$\int_a^b f(x)\,dx = \left[F(x)\right]_a^b, \text{ where } \int f(x)\,dx = F(x) + c.$$

So, for example, $\displaystyle\int_{-1}^{2} 4x^3\,dx = \left[x^4\right]_{-1}^{2}$

$$= (2^4) - ((-1)^4)$$
$$= (16) - (1)$$
$$= 15$$

Checkpoint

$x^4 \xrightarrow{\text{differentiates to}} 4x^3$

Common error

$(-1)^4$ is **not** -1.

Use brackets to avoid this error.

Taking it further

You will use integration to find the exact area of a region between a curve and the x-axis. You will also need integration to solve **differential equations** which model the behaviour of real-world events such as the growth of a virus.

Checkpoint

Verify directly that $\displaystyle\int_0^{1/2} f(x)\,dx = 0$.

9.2 Applying integration

What you should already know:

- how to find the area of a trapezium.
- how to evaluate a definite integral such as $\displaystyle\int_2^4 x^2 - 3x + 1\,dx$.
- how to find the coordinates of the points where a line and curve intersect.

In this section you will learn:

- how to calculate the exact area of a region between a curve and the x-axis.

At GCSE, you estimated the area beneath a graph by using triangles or trapeziums.

GCSE Example 3

The graph shows the speed, in metres per second, of a car t seconds after it pulls away from a set of traffic lights.

a Calculate an estimate for the distance travelled by the car in the first 9 seconds.

Use 3 strips of equal width.

b Use your answer to part **a** to find an estimate for the average speed of the car during the first 9 seconds.

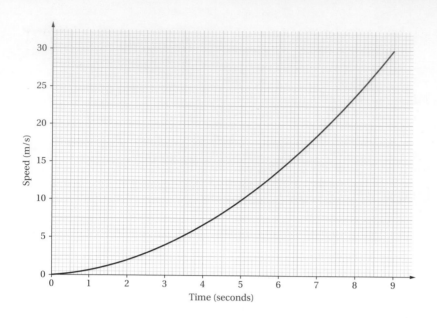

Working

a The area under a speed–time graph measures the distance travelled.

You can divide up the area under this graph using a triangle and two trapeziums, as shown.

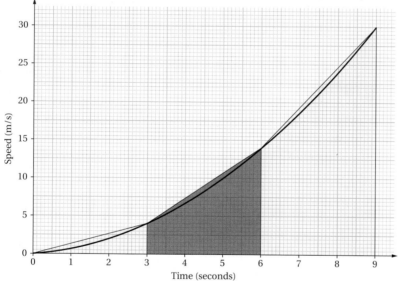

The area of the triangle is $\frac{1}{2} \times 3 \times 4 = 6$.

The area of the first trapezium is $\frac{1}{2} \times (4 + 14) \times 3 = 27$.

The area of the second trapezium is $\frac{1}{2} \times (14 + 30) \times 3 = 66$.

Add these areas together: $6 + 27 + 66 = 99$

Handy hint

Use simple ratios to work out each strip width. For 3 strips of equal width to cover a 9 second period, each strip must be 3 wide.

The distance travelled by the car in the first 9 seconds is approximately 99 metres. This is an overestimate for the exact distance travelled in this time, because the lines joining the tops of the three shapes lie entirely above the curve.

b Average speed $= \dfrac{\text{total distance}}{\text{total time}}$

$\approx \dfrac{99}{9} = 11$ metres per second.

At A Level, you will need to be able to find the **exact** area of a region between a curve and the x-axis.

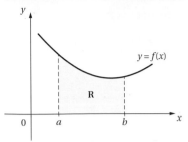

Handy hint

'Bounded' means 'trapped between'.

Key point

The diagram above shows the curve with equation $y = f(x)$.

R is the region bounded by this curve and the x-axis between $x - a$ and $x - b$.

The exact area of $\mathbf{R} = \int_a^b f(x)\, dx$ provided the whole of **R** lies above the x-axis.

As you progress through your course, you will learn why integration gives a measure of area. You will also learn how to deal with regions which lie below the x-axis.

The equation of the curve shown in the diagram in Example 3 is $y = \frac{1}{3}x^2 + \frac{1}{3}x$.

Using integration, the exact area bounded by this curve between $x - 0$ and $x - 9$.

$$= \int_0^9 \frac{1}{3}x^2 + \frac{1}{3}x\, dx$$

$$= \left[\frac{1}{9}x^3 + \frac{1}{6}x^2 \right]_0^9$$

$$= \left(\frac{1}{9}(9)^3 + \frac{1}{6}(9)^2 \right) - \left(\frac{1}{9}(0)^3 + \frac{1}{6}(0)^2 \right)$$

$$= (94.5) - (0)$$

$$= 94.5$$

So the exact distance travelled by the car in the first 9 seconds is 94.5 metres.

This confirms that the approximate answer 99 metres found in part **a** of Example 3 is an overestimate for the exact distance travelled.

At A Level you often need to calculate the area of a region which is bounded between a line and a curve, or between two curves.

Handy hint

You can use the curve equation to find the side lengths of the trapeziums in Example 3.

For example, when $x - 9$,

$y = \frac{1}{3}(9)^2 + \frac{1}{3}(9) - 30$

where 30 is the longer of the two side lengths of the second trapezium.

A Level Example 4

The diagram shows the curve with equation $y = 3 + 6x - x^2$ and the line with equation $y = x + 7$.

This curve and line intersect at points A and B, with x-coordinates 1 and 4, respectively.

Find the area of the shaded region $\mathbf{R}$ shown in the diagram.

Working

Between the x-axis, $x = 1$ and $x = 4$, the area under the curve minus the area under the line equals the area of region $\mathbf{R}$.

So the area of $\mathbf{R} = \int_1^4 3 + 6x - x^2 \, dx \; - \; \int_1^4 x + 7 \, dx$.

Work out each integral:

$$\int_1^4 (3 + 6x - x^2) \, dx = \left[3x + 3x^2 - \frac{1}{3}x^3 \right]_1^4$$

$$= \left(3(4) + 3(4)^2 - \frac{1}{3}(4)^3 \right) - \left(3(1) + 3(1)^2 - \frac{1}{3}(1)^3 \right)$$

$$= \left(\frac{116}{3} \right) - \left(\frac{17}{3} \right)$$

$$= 33$$

$$\int_1^4 x + 7 \, dx = \left[\frac{1}{2}x^2 + 7x \right]_1^4$$

$$= \left(\frac{1}{2}(4)^2 + 7(4) \right) - \left(\frac{1}{2}(1)^2 + 7(1) \right)$$

$$= (36) - \left(\frac{15}{2} \right)$$

$$= 28.5$$

So the area of $\mathbf{R} = 33 - 28.5$

$$= 4.5$$

In this example, you can combine the integrands so that a single integration is required to find the area of $\mathbf{R}$.

So area of $\mathbf{R}$
$$= \int_1^4 3 + 6x - x^2 \, dx - \int_1^4 x + 7 \, dx$$

$$= \int_1^4 (3 + 6x - x^2) - (x + 7) \, dx$$

$$= \int_1^4 -4 + 5x - x^2 \, dx$$

Take care – you can only combine the integrands in this way because the integrals use the same pair of upper and lower limits.

> **Taking it further** >>>>>

You will need to understand how to find the area of a region part (or all) of which lies below the x-axis.

A Level Alert!

In a harder question, you would need to find the coordinates of points A and B by solving simultaneous equations – see Section 5.3.

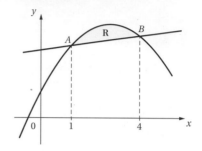

Handy hint

'Bounded' means 'trapped between'.

Checkpoint

Verify the answer 28.5 by working out the area of the trapezium **T**. Use the equation of the line to find the lengths of the vertical sides of **T**.

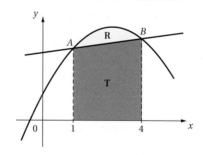

Checkpoint

Verify that $\int_1^4 -4 + 5x - x^2 \, dx = 4.5$.

1 Practice: Surds and indices

1.1 Surds

Unless you are told otherwise, do not use a calculator for these questions.

1 Simplify these.

a $3\sqrt{5} + 4\sqrt{5}$ **b** $2\sqrt{3} \times 3\sqrt{2}$

c $\dfrac{4\sqrt{10}}{2\sqrt{5}}$ **d** $(2\sqrt{5})^2$

e $\left(-2\sqrt{2}\right)^3$ **f** $8\sqrt{\dfrac{7}{16}}$

2 Express these in simplified surd form.

a $\sqrt{45} + \sqrt{20}$ **b** $\sqrt{32} - \sqrt{18}$

c $2\sqrt{48} - 4\sqrt{27}$

3 By simplifying each surd, find the value of $\dfrac{\sqrt{50} + \sqrt{32}}{\sqrt{72} - \sqrt{18}}$

4 Simplify these expressions.

a $\left(1 + \sqrt{2}\right)\left(2 + \sqrt{2}\right)$

b $\left(4 + \sqrt{3}\right)\left(1 - \sqrt{3}\right)$

c $\left(\sqrt{2} + 3\right)\left(2\sqrt{2} - 1\right)$

d $\left(2 + \sqrt{6}\right)^2$

e $\left(5 - 2\sqrt{3}\right)\left(4 + 3\sqrt{3}\right)$

f $\left(\sqrt{2} + \sqrt{3}\right)^2$

5 Express these fractions in the form $a + b\sqrt{3}$, where a and b are integers.

a $\dfrac{1}{2 + \sqrt{3}}$ **b** $\dfrac{12}{3 - \sqrt{3}}$

c $\dfrac{4\sqrt{3}}{\sqrt{3} + 1}$

6 Simplify these fractions.

a $\dfrac{5 + \sqrt{7}}{3 - \sqrt{7}}$ **b** $\dfrac{4 - \sqrt{3}}{2 - \sqrt{3}}$

c $\dfrac{3 + \sqrt{3}}{3 - 2\sqrt{3}}$

7 Show that $\dfrac{\sqrt{24} - 6}{3 - \sqrt{6}}$ is an integer, stating its value.

8 Given that $D = b^2 - 4ac$:

a find the value of $\sqrt{D}$ when

> *Handy hint*
> $\sqrt{D}$ means the *positive* square root of D.

   i $a = 2, b = 4, c = 1$

   ii $a = 3, b = -1, c = -2$

   Give each answer in simplified surd form.

b Explain why $\sqrt{D}$ does not have a real value when $a = b = c$ where $b \neq 0$.

9 Express these in simplified surd form.

a $\sqrt{75}$ **b** $\sqrt{180}$

c $\sqrt{192}$ **d** $\sqrt{150} + \sqrt{96}$

10 ABC is a right-angled triangle. $AB = 4 + 2\sqrt{3}$, $AC = 4 - 2\sqrt{3}$.

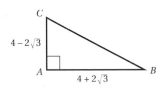

a Find the area of this triangle.

b Find the exact perimeter of this triangle. Give your answer in the form $a + b\sqrt{14}$, where a and b are integers to be stated.

(PS) 11 a Find the greatest of these numbers. You may use a calculator if you wish.

$$1 + \sqrt{3}, \ 2 + 2\sqrt{3}, \ 3 + \sqrt{3}$$

b Show that these three numbers are sides of a right-angled triangle.

c Find the area of this triangle, giving your answer in the form $a + b\sqrt{3}$, where a and b are integers to be stated.

(PS) 12 PQR is a right-angled triangle. $PQ = 5 + \sqrt{5}$, $PR = 3 + 3\sqrt{5}$.

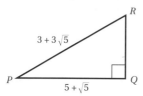

a Expand $\left(1 + \sqrt{5}\right)^2$.

b Show that $QR^2 = 24 + 8\sqrt{5}$.

c Show that the area and perimeter of this triangle are numerically equal.

1.2 Indices

1 Express each of these in the form 2^n where n is an integer.

a $2^3 \times 2^4$ **b** $(2^3)^3$

c 4^5 **d** $(2^4 \times 4^2)^3$

2 Express these numbers in the required form.

a 4^{-2} in the form 2^n **b** 2^{-6} in the form 4^n

c $8^{\frac{4}{3}}$ in the form 4^n **d** $27^{\frac{5}{3}}$ in the form 9^n

e $16^{-\frac{1}{2}}$ in the form 8^n **f** $64^{-\frac{4}{3}}$ in the form 16^n

3 By writing 16 as a power of 2, or otherwise, solve the equation $16^x = 32$.

4 Solve these equations.

a $8^x = 16$ **b** $16^x = 64$

c $9^x \times 3^x = 9$ **d** $\dfrac{8^x}{4^{x+1}} = 32$

5 Express these terms in the form ax^n where a is a real number.

a $\dfrac{4x}{2x^2}$ **b** $\dfrac{1}{2x^3}$

c $3x\sqrt{x}$ **d** $\dfrac{\sqrt[3]{x^2}}{4}$

e $\dfrac{2}{\sqrt{x}}$ **f** $\dfrac{3x}{\sqrt[3]{x}}$

g $\dfrac{3\sqrt{x^3}}{6x^2}$ **h** $\dfrac{10x}{\sqrt[4]{x^3}}$

6 Determine whether each of these statements is true or false. Use rules of indices to prove those which you think are true. For those statements that you think are false, give an example to show that it is incorrect.

a $a^n \times a^n = a^{2n}$ for all numbers a and positive integers n

b $a^n \times b^n = ab^n$ for all numbers a and b and positive integers n

c $a^{mn} = a^m \times a^n$ for all numbers a and positive integers m and n

d $a^n \times a^{-n} = 1$ for all non-zero numbers a and positive integers n

e $(a^n)^n = a^{2n}$ for all numbers a and positive integers n

7 a Express $\dfrac{3x^3 + 2}{x^2}$ in the form $ax + bx^n$, where a, b and n are constants.

b Express $\dfrac{2x^2 - 3x + 1}{2x^2}$ in the form $a + bx^{-1} + cx^{-2}$, where a, b and c are constants.

8 Express these as sums of powers of x.

a $\dfrac{(2x + 1)(x - 1)}{x}$

b $\dfrac{(3x + 2)^2}{x^3}$

c $\dfrac{x^2 + 3x - 6}{\sqrt{x}}$

d $\dfrac{(2 + \sqrt{x})^2}{x^2}$

9 A curve C has equation $y = \dfrac{(3x + 2)(2x + 3)}{x^2}$
(PS) where $x \neq 0$.

a Express y in the form $a + bx^{-1} + cx^{-2}$, where a, b and c are constants.

b Explain why, as x increases and is positive, the value of y approaches, but never equals, 6.

c Is there a point on this curve with y-coordinate 6?

2 Practice: Algebra 1

2.1 Basic algebra

1 Expand and then simplify these expressions.

   a $2(a + 3) + 3(a - 1)$

   b $3(b + 2) - 4(2b - 3)$

   c $4(a + 2b) + 2(3a - 4b)$

   d $a(2a + b) - b(a - 3b)$

2 Factorise fully these expressions.

   a $2x^3y + xy^2$ **b** $10x^3y^2 - 4x^2y^3$

   c $3x^4y^2z + 6x^3yz^2$ **d** $12x^4y^2 + 6x^2y^2 - 9xy$

3 Expand these expressions. Fully factorise answers where appropriate.

   a $(2a^2b)^2$ **b** $(3ab^2)^3 + (3a^2b)^2$

   c $(4a^2b^2)^2 - (2ab^3)^2$

4 Rearrange these equations to make the variable shown in square brackets the subject.

   a $P = 3(Q + 4)$ $[Q]$

   b $A = \frac{1}{2}(3B - 1)$ $[B]$

   c $R + T = 3(T - 1)$ $[T]$

   d $2(C - D) = 5(1 + 2D)$ $[D]$

   e $U = \frac{1}{3}\sqrt{V + 2}$ $[V]$

   f $M = \frac{\pi}{2}(N - 1)^3$ $[N]$

5 The volume V of a sphere with radius r is $V = \frac{4}{3}\pi r^3$.

   a Re-arrange this formula to make r the subject.

   b Find the radius of a sphere with volume 36π cm³.

6 The diagram shows a square $PQRS$ of side length $2x$ cm. A quarter circle, centre P and radius $2x$ cm, is inscribed inside the square.

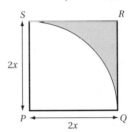

   a Show that the area A of the shaded shape is given by the formula $A = 4x^2 - \pi x^2$.

   b Make x the subject of this formula.

   c Show that the perimeter of the shaded shape is given by the expression

$$(4 + \pi)\sqrt{\frac{A}{4 - \pi}}.$$

7 The diagram shows a right-angled triangle ABC, where $AB = m + 1$, $BC = m - 1$ and $AC = n$.

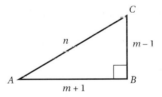

   a If $t = \tan \hat{A}$ show that $m = \frac{1 + t}{1 - t}$.

   b Find an expression for m in terms of n.

8 Rearrange these formulae to make x the subject.

   a $y = (x + 3)^2$

   b $y = 4(x - 1)^2 - 1$

   c $y = \frac{(2x - 5)^2}{3}$

9 $P = \dfrac{2Q + 3}{Q}$

    a Show that $P = 2 + \dfrac{3}{Q}$.

    b Hence, or otherwise, make Q the subject of the formula $P = \dfrac{2Q + 3}{Q}$.

10 Make the letter indicated in square brackets the subject of these formulae.

    a $A = \dfrac{B - 2}{B}$ [B]

    b $C = \dfrac{D^2 + 4}{D^2}$ [D]

    c $E = \dfrac{5 - 4F^3}{F^3}$ [F]

11 Make the letter indicated in square brackets the subject of these formulae.

    a $A = \dfrac{B}{B - 2}$ [B]

    b $C = \dfrac{D + 2}{2D + 3}$ [D]

    c $E = \dfrac{F^2 + 3}{F^2 + 1}$ [F]

12 Simplify these fractions.

    a $\dfrac{x^2 + 3x}{x}$ **b** $\dfrac{2x^4 + 4x^2}{x^2}$

    c $\dfrac{3x^2 - 3x}{x - 1}$ **d** $\dfrac{x^2 - 2x^3}{2x - 1}$

13 Express these fractions in the required form. State the value of each constant.

    a $\dfrac{3x^3 + 4x^2 + 6x}{3x}$ in the form $Ax^2 + Bx + C$ for constants A, B and C.

    b $\dfrac{4x^3 - 3x}{2x^2}$ in the form $Ax - Bx^{-1}$ for constants A and B.

    c $\dfrac{6x^4 + 9x^2 + 1}{3x^2}$ in the form $Ax^2 + Bx^{-2} + C$ for constants A, B and C.

    d $\dfrac{12x^4 - 4x^3}{3x - 1}$ in for the form Ax^n for constants A and n.

2.2 Solving linear equations

1 Solve these equations.

    a $3(a + 2) = 21$ **b** $8 + \frac{1}{2}(b - 2) = 11$

    c $5(11 - 2c) = 25$ **d** $\dfrac{19 - 3d}{4} = 7$

2 Solve these equations.

    a $4a - 3 = 2a + 7$ **b** $6b + 1 = 3 - 4b$

    c $5(c - 2) = 8c + 2$

3 Solve these equations.

    a $\dfrac{3x}{4} + 7 = 10$ **b** $\dfrac{x}{2} + \dfrac{x}{4} = 6$

    c $\dfrac{x}{2} + \dfrac{2x}{3} = 7$

4 By expressing these as linear equations, find the value of x.

    a $\dfrac{12}{x + 1} = 3$ **b** $\dfrac{8x}{2x + 3} = 1$

    c $\dfrac{3}{x} + \dfrac{1}{2x} = 14$

5 For the equation $5x - 2y = k$, where k is a constant, it is known that $y = 3$ when $x = 6$.

    a Show that $k = 24$.

    b Hence find the value of x when $y = 8$.

6 Solve these simultaneous equations.

    a $3x + 2y = 12$ **b** $4x - 3y = 31$
        $2x + 3y = 13$ $5x + 2y = 33$

    c $8x - 6y = 13$
        $3x - 5y = 9$

7 The diagram shows a rectangle $ABCD$. $AB = 2x + 1$, $AD = 3x - 2$, where all dimensions are in centimetres.

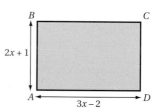

    a Find an expression for the perimeter of this rectangle. Simplify your answer as far as possible.

The perimeter of this rectangle is 33 cm.

    b Find the area of this rectangle.

8 The shape in the diagram shows a rectangle $ABCD$ supporting a semi-circle, where $AB = (3x - 1)$ cm and the semi-circle has radius x cm. The centre of the semi-circle is the mid-point of BC.

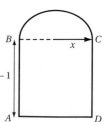

a Show that the perimeter P of the shape is given by the formula $P = (\pi + 8)x - 2$.

b Given that the perimeter of the shape is 555 cm, use a calculator to find the area of this shape. Give your answer to 3 significant figures.

9 By multiplying each of these equations by a suitable number, or otherwise, find the value of x for which:

a $\dfrac{x+3}{4} + \dfrac{x}{2} = 6$ **b** $\dfrac{2x-1}{3} - \dfrac{x+4}{6} = 5$

c $\dfrac{3x+1}{2} + \dfrac{x}{3} = 6$

2.3 Forming expressions

Unless you are told otherwise, assume all lengths are in centimetres.

1 The diagram shows a rectangle $ABCD$. Point E is the mid-point of BC.

$AB = x$, $AD = 2x + 4$

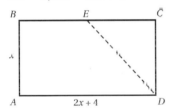

a Find an expression in terms of x for the perimeter of this rectangle.

b Show that the area of the trapezium $ABED$ is given by the formula

Area $= \dfrac{3}{2}x(x + 2)$.

2 The diagram shows the shape formed when a square of side length x is removed from the rectangle $ABCD$, where $AD = 2x + 5$ and $CD = 3x$.

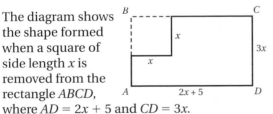

a Find, in factorised form, an expression for

  i the perimeter of the shape

  ii the area of the shape.

The area of the removed square is 49 cm².

b Find the area of the shape.

3 The diagram shows two circles with a common centre. The radius of the smaller circle is x cm. The (shortest) gap between the two circles is 3 cm.

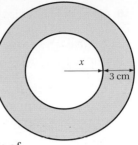

a Find an expression for the circumference of the larger circle. Leave π in your answer.

b Show that the area of the shaded region is given by the formula

Area $= 3\pi(2x + 3)$.

4 The diagram shows a circle with radius r and centre O. Points A and B on the circle are such that triangle AOB is right-angled.

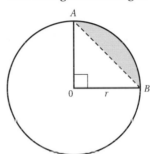

> *Handy hint*
>
> A sector of a circle looks like a slice of pizza!

a Show that the perimeter P of the sector OAB is given by the formula

$P = \dfrac{1}{2}r(4 + \pi)$.

b Find an expression in terms of r for the area of the segment shaded in the diagram. Factorise your answer as far as possible.

5 The diagram shows a rectangle $ABCD$.

$AB = x + y$,
$BC = x - y$,
where $x > y$.

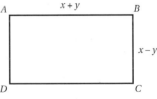

a Find an expression for the perimeter P of this rectangle.

The area of this rectangle is equal to the area of a square with side length y.

b Use this information to show that $x = ky$, stating the exact value of k.

6 An area of land is fenced off using some barbed wire and a wall. In the diagram the wire is represented by the edges *AB*, *BC* and *CD*. The side *AD* represents the wall, where *ABCD* is a rectangle.

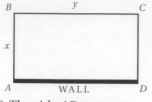

The total length of barbed wire used is 24 metres.

a Express this information as an equation involving x and y.

b Hence show that the area of this enclosure is given by the formula: Area $= 2x(12 - x)$.

c Find the enclosed area in the case when *ABCD* is a square.

7 The diagram shows a cuboid with dimensions x, $2x$ and $3x$.

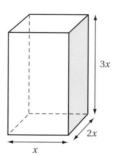

a Find an expression for the volume V of this cuboid. Simplify your answer as far as possible.

b Show that the external surface area S of the cuboid is given by the formula $S = 22x^2$.

c Express the area of the side shaded in the diagram as a fraction of the external surface area. Give your answer in its lowest terms.

The volume of this cuboid is 48 cm³.

d Find the external surface area of this cuboid.

8 From a rectangle, four squares of side length x cm are cut from each corner.

Diagram 1 shows the net of the remaining shape.

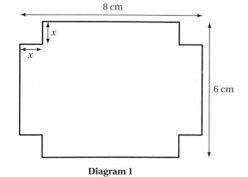

Diagram 1

a Find an expression in terms of x for the area of this net.

The sides of this net are folded at the corners to form a tray with an open top (see Diagram 2).

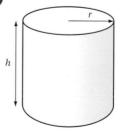

Diagram 2

b Show that the volume V of this tray is given by the formula $V = 4x(4 - x)(3 - x)$.

The total surface area of this tray is 78 cm².

c Find the volume of this tray.

>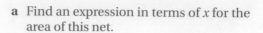
> *Handy hint*
> You must include both the exterior and the interior surfaces of the tray.

9 The diagram shows a cuboid with base dimensions x cm by y cm. The cuboid has height 4 cm.

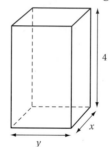

a Find an expression involving x and y for the external surface area S of this cuboid.

The volume of this cuboid is 16 cm³.

b Use this information to show that $xy = 4$.

c Find an expression for S in terms of x only.

10 The diagram shows a cylinder with radius r and height h.

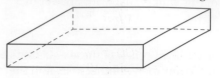

a Write down the volume V of this cylinder in terms of r and h.

It is given that the volume of this cylinder is 9π cm³.

A straight metal rod, which is the longest that can be placed in the cylinder, has length L.

b Show that $L = \sqrt{\dfrac{36}{h} + h^2}$.

 11 The rectangle $ABCD$ shown in Diagram 1 is curled so that the side AB meets the side CD to make the hollow cylinder with height 4 cm and radius r shown in Diagram 2.

$AB = 4$ cm and $BC = L$ cm.

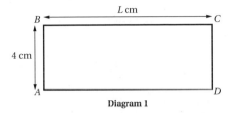

Diagram 1

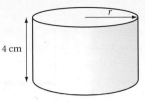

Diagram 2

a Show that the volume V of this cylinder is given by the formula

$$V = \frac{L^2}{\pi}.$$

A circular lid and a base each of radius r are added to the cylinder in Diagram 2.

b Show that the surface area S of this closed cylinder is given by the formula

$$S = \frac{L}{2\pi}(8\pi + L).$$

3 Practice: Coordinate geometry 1

3.1 Straight-line graphs

1 Find the gradient m and the y-intercept c of each of these lines.

   a $y = 3x + 6$ **b** $y = 2 - 4x$

   c $y = \dfrac{4x - 5}{2}$ **d** $y = -\dfrac{1}{3}(3 + 4x)$

Sketch, on separate diagrams, the lines with these equations. Label the points where the line crosses each axis with their coordinates.

2 By making y the subject of each equation, find the gradient m and the y-intercept c of these lines.

   a $y - 2x + 1 = 0$ **b** $2y - 3x = 2$

   c $4x - 3y = 1$ **d** $\dfrac{y}{4} + \dfrac{x}{2} = 3$

3 Show that these pairs of lines are parallel.

   a $y = 2x - 4$ **b** $2y - 3x = 1$

      $y - 2x + 3 = 0$ $y = \dfrac{4 + 3x}{2}$

   c $2x + 4y - 3 = 0$

      $2y + x + 1 = 0$

4 Show that these pairs of lines are perpendicular.

   a $y = 3x + 4$ **b** $2y + 3x = 0$

      $3y + x = 3$ $3y - 2x = 2$

   c $y = \dfrac{11 - 5x}{3}$

      $3x - 5y + 1 = 0$

5 **a** Sketch, on the **same** diagram, the line $y - 2x = 1$ and the line $2y - 6x + 1 = 0$.

   b Find the distance between the y-intercepts of these graphs.

6 **a** Sketch, on the **same** diagram, the line $y - 3x + 4 = 0$ and the line $3y + x = 6$.

   b Find the distance between the x-intercepts of these graphs.

7 Express the equations of these lines in the form $ay + bx + c = 0$, where a, b and c are integers.

   a $y = -\dfrac{1}{2}x - \dfrac{3}{2}$ **b** $y = \dfrac{1}{3} - \dfrac{2x}{3}$

   c $y = -\dfrac{3}{4}x + \dfrac{1}{2}$ **d** $y = \dfrac{2}{3}x - \dfrac{5}{2}$

8 The diagram shows a sketch of the lines A, B, C and D. The lines have equations (1), (2), (3) and (4).

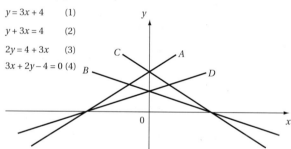

$y = 3x + 4$ (1)

$y + 3x = 4$ (2)

$2y = 4 + 3x$ (3)

$3x + 2y - 4 = 0$ (4)

Match each line with its equation.

9 The height h (in cm) of a sunflower t days after being planted is modelled by the equation $h = 3.5t + 10$.

   a Sketch the graph of h against t for $t \geq 0$.

   b Explain, in context, what each constant in the equation $h = 3.5t + 10$ represents.

When fully grown the sunflower is expected to be 5 metres tall.

   c According to this model, how many weeks from being planted does it take for the sunflower to reach its full height?

   d Give one reason why this model may not be appropriate.

10 The speed v (in metres per second) of an athlete t seconds after crossing the finish line in a 100 metre race is modelled by the equation $v = 7 - 1.4t$.

  a Sketch the graph of v against t for $t \geq 0$, labelling the axis-crossing points with their coordinates.

> *Handy hint*
>
> If necessary revise speed–time graphs from GCSE.

  b Describe, in context:

    i what each axis-crossing point represents

    ii what the gradient represents.

  c According to this model calculate the total distance the athlete runs from the start of the race until he stops.

(PS) 11 The line L has equation $ay + bx = 12$, where a and b are positive constants.

L crosses the y-axis at point A and crosses the x-axis at point B.

  a Sketch the graph of L, labelling points A and B with their coordinates in terms of a and b, respectively.

Triangle OAB has area 9 square units.

  b Show that $ab = 8$.

The gradient of L is -2.

  c Find the value of a and the value of b.

  d Show that the length of the shortest line from O to L is $\dfrac{6\sqrt{5}}{5}$.

3.2 Finding the equation of a line

1 a Find the equations of the lines which pass through these points. Give each answer in the form $y = mx + c$.

    i $A(3, 4)$ and $B(6, 10)$

    ii $A(1, -3)$ and $B(3, 7)$

    iii $A(-1, 5)$ and $B(3, 7)$

    iv $A(-4, -1)$ and $B(-6, 5)$

  b One of your answers to part **a** does not involve integers. Write this equation in the form $ay + bx = c$ where a, b and c are integers.

2 The sketch shows the line L which passes through the points $A(0, 3)$ and $B(2, 2)$. This line crosses the x-axis at point C.

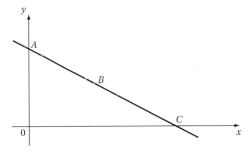

  a Find the gradient of L.

  b Find the equation of L. Give your answer in the form $ay + bx = c$ where a, b and c are integers.

  c Find the coordinates of point C.

3 Find the equations of these lines.

  a The line with gradient 4 which passes through the point $A(2, 3)$.

  b The horizontal line which passes through the point $C(35, -7)$.

  c The line with gradient $-\dfrac{3}{2}$ which passes through the point $B(4, -2)$.

    Give your answer for this line in the form $ay + bx = c$ where a, b and c are integers.

4 a A line passes through the points $A(1, 5)$ and $B(3, p)$, where p is a constant. Given that the gradient of this line is 4, find the value of p.

  b A line passes through the points $A(q, 12)$ and $B(6, q)$, where q is a constant. Given that the gradient of this line is -7, find the value of q.

  c A line passes through the points $E(r, r + 1)$ and $F(8, 0)$, where r is a constant.

    Given that the gradient of this line is $-\dfrac{1}{2}$, find the value of r.

5 The line L passes through the points $P(-4, -3)$ and $Q(4, 9)$. This line crosses the y-axis at point A and the x-axis at point B.

> *Handy hint*
>
> Sketch the line L.

  a Find an equation for L.

  b Find the area of triangle OAB, where O is the origin.

6 A line passes through the points $S(3, -2)$ and $T(12, -14)$. This line crosses the y-axis at point A and the x-axis at point B.

Handy hint
Sketch the line.

Show that the distance $AB = \dfrac{5}{2}$.

7 A line has equation $y = mx - 3$ where m is a constant. The point $A(-5, 7)$ lies on this line.

   a Find the value of m.

   b Determine whether or not the point $B(-7, 10)$ lies on this line.

(PS) 8 The line L has equation $2y - 4x + k = 0$ where k is a constant.

   The point $A\left(\dfrac{5}{2}, \dfrac{1}{2}\right)$ lies on L.

   a Show that $k = 9$.

   b Find the y-intercept of L.

   c Find the area of the triangle formed by this line and the two coordinate axes.

9 The line L has equation $y - 3x + 1 = 0$. The points $A(3, 8)$ and $B(-1, k)$, where k is a constant, lie on L.

   a Show that $k = -4$.

   b Find the equation of the perpendicular bisector of AB. Give your answer in the form $ay + bx = c$, for integers a, b and c.

(PS) 10 The line L passes through the points $P(1, 4k)$ and $Q(k, 4)$, where k is a constant, $k > 1$.

   a Show that the gradient of L is -4.

   b Find the equation of L, giving your answer in terms of k.

   Point R is where L crosses the x-axis.

   c Find the value of k such that $PQ = QR$.

11 The line L_1 passes through the points $A(3, 2)$ and $B(9, 0)$. The line L_2 is parallel to L_1 and passes through the point $C(7, 4)$.

   a Find an equation for L_2.

   b Show that the perpendicular bisector of AB passes through point C.

   c Hence, show that the perpendicular distance between L_1 and L_2 is $\sqrt{10}$ units.

12 The line L_1 passes through the points $A(-3, -1)$ and $B\left(1, \dfrac{5}{3}\right)$.

   a Show that the equation of L_1 can be written as $3y = 2x + 3$.

   The line L_2 has equation $3x + 2y = 28$.

   b Sketch, on a single diagram, the lines L_1 and L_2.

   c Verify that point $P(6, 5)$ lies on both L_1 and L_2.

Handy hint
For part **c** use the coordinates of P to confirm that P lies on each line.

   d Find the area of the triangle formed by these lines and the y-axis.

3.3 Mid-points and distances

1 Find the coordinates of the mid-point of the line AB.

   a $A(2, 5)$, $B(10, 3)$

   b $A(5, -1)$, $B(-1, 7)$

   c $A(-6, 11)$, $B(3, 4)$

   d $A\left(\dfrac{3}{2}, \dfrac{5}{3}\right)$, $B\left(\dfrac{5}{2}, \dfrac{1}{6}\right)$

2 Find, in terms of the constant k, the coordinates of the mid-point of AB. Simplify each answer as far as possible.

   a $A(2, 4)$, $B(k, 2k)$

   b $A(-2k, 5)$, $B(0, 2k+1)$

   c $A(3k, 2-k)$, $B(k, 5k)$

3 The mid-point of AB is the point $C(2, 3)$. If A has coordinates $(1, -2)$ find the coordinates of B.

4 Points $A(-4, 4)$ and B are such that the mid-point of AB is the point $C(-3, 7)$.

   Find the coordinates of the point D such that B is the mid-point of CD.

5 Points $A(p, 3)$ and $B(14, q)$, where p and q are constants, are such that the mid-point of AB is the point $C(8, 11)$. Point D is the mid-point of AC.

a Show that $p = 2$ and find the value of q.

b Find the coordinates of D.

It is given that the distance $AD = 5$.

c Find the distance DB.

(PS) 6 The vertices of the square $ABCD$ have coordinates $A(1, 3)$, $B(-3, 7)$, $C(p, q)$ and $D(5, 7)$, where p and q are constants.

The diagonals of any square bisect each other. Find the coordinates of C.

7 Find the distance AB for these points. Give answers in simplified surd form where appropriate.

  a $A(1, 4)$, $B(6, 2)$

  b $A(2, -2)$, $B(5, 7)$

  c $A(-3, -1)$, $B(-5, 9)$

  d $A\left(-\dfrac{1}{2}, \dfrac{3}{4}\right)$, $B\left(\dfrac{1}{2}, \dfrac{3}{2}\right)$

8 Given the points $A(3, 4)$, $B(6, -1)$ and $C(-2, 7)$, prove that the triangle ABC is isosceles, but **not** equilateral.

9 A circle has diameter AB, where $A(4, 1)$ and $B(0, 2)$.

  a Find the coordinates of the centre of this circle.

  b Show that the radius of this circle is $\dfrac{5}{2}$ units.

10 The point $C(2, -3)$ is the centre of a circle. The point $A(7, 9)$ lies on this circle.

  a Show that the radius of this circle is 13 units.

  b Find the coordinates of the point B such that AB is a diameter of this circle.

(PS) 11 The diagram shows the points $A(0, 3)$, $B(3, 9)$ and $C(5, 8)$

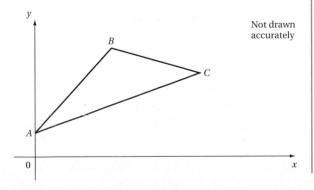

Not drawn accurately

a Show that $AB^2 + BC^2 = AC^2$

b What information about the triangle ABC does the result of part **a** give?

c Find the area of the circle which passes through points A, B and C. Leave π in your answer.

12 The points $A(p, 3)$ and $B(1, 6)$, where p is a **(PS)** positive constant, are such that the distance $AB = p$. C is the midpoint of AB.

a Show that $p = 5$.

b Find the circumference of the circle which has centre C and passes through point A. Leave π in your answer.

3.4 Intersections of lines

1 i On separate diagrams, sketch the pairs of lines with these equations.

ii Use algebra to find the coordinates of the points where each pair of lines intersect. Give answers as top-heavy fractions where appropriate.

  a $y = 4x \quad 9$
  $y = 2x + 3$

  b $y = \dfrac{x + 3}{2}$
  $y = 2 - \dfrac{1}{2}x$

  c $y = \dfrac{1}{3}x + \dfrac{3}{2}$
  $y = \dfrac{2}{3} - \dfrac{1}{2}x$

2 Use algebra to find the coordinates of the points where these lines intersect.

  a $2y + 3x - 5 = 0$
  $y = 4 - 2x$

  b $3y - 4x = 8$
  $y = \dfrac{x + 2}{3}$

  c $2y = x - 4$
  $2x - 5y - 12 = 0$

3 Line L_1 has equation $8y + 6x = 5$. Line L_2 has equation $y = 2 - \dfrac{3}{4}x$.

  a Use algebra to show that there is no solution to the simultaneous equations $8y + 6x = 5$ and $y = 2 - \dfrac{3}{4}x$.

  b What geometrical information does the result of part **a** give about the lines L_1 and L_2?

c Sketch the lines L_1 and L_2 on a single diagram.

4 By expressing each equation in the form $y = mx + c$, for constants m and c, determine whether these pairs of lines intersect. For those that do, find the coordinates of the point of intersection.

a $4x + 3y = 6$
$3y - 4x + 18 = 0$

b $6x - 10y + 9 = 0$
$5y - 3x - 10 = 0$

c $6y + x = 2$
$3y = 1 + \frac{x}{2}$

5 i By solving these simultaneous equations, find the coordinates of the points where these lines intersect.

ii Sketch each pair of lines on a separate diagram.

a $2y - x = 5$
$2y + x = 9$

b $3y + x = 12$
$2y + 3x = 8$

c $2y - 3x + 9 = 0$
$5y - 2x = 5$

6 The diagram shows the lines L_1 and L_2. L_1 crosses the x-axis at the point A and has equation $y = 3x - 3$. L_2 crosses the x-axis at the point B and has equation $y = 7 - 2x$. These lines intersect at point C.

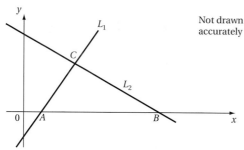

a Find the x-coordinate of point A and the x-coordinate of point B.

b Find the area of triangle ABC.

7 The diagram shows the line L_1 with equation $2y - x = 6$ and the line L_2 with equation $3y + 2x = 16$. These lines intersect at point C.

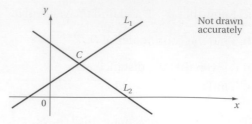

Find the area of the triangle formed by these lines and:

a the y-axis

b the x-axis.

8 The line L_1 has equation $y = 6x - 3$. The line L_2 has equation $9y - 12x = 1$.

a Find the coordinates of the point A where these lines intersect.

b Find the equation of the line perpendicular to L_2 which passes through point A. Give your answer in the form $ay + bx = c$ for integers a, b and c.

9 The diagram shows triangle ABC formed by the intersection of three lines. Point A has coordinates $(10, 7)$. The line L_1 passes through A and C and has gradient $\frac{1}{4}$.

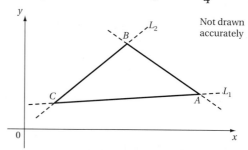

a Show that an equation for L_1 is $4y - x = 18$.

Line L_2 passes through the points B and C and has equation $2y - 3x = 4$.

b Find the coordinates of the point C.

The y-intercept of the line which passes through the points A and B is 17.

c Find the distance BC, giving your answer in simplified surd form.

4 Practice: Algebra 2

4.1 Solving a quadratic equation by factorising

1 Solve these equations using factorisation.

a $x^2 - 8x + 15 = 0$

b $x^2 + 5x - 14 = 0$

c $x^2 - 6x + 9 = 0$

d $2x^2 + x - 10 = 0$

e $2x^2 - 11x + 5 = 0$

f $3x^2 + 5x - 12 = 0$

2 Rearrange these equations into a suitable form and then solve them using factorisation.

a $x^2 + 3x - 18$

b $2x^2 + 6 = 7x$

c $x(x - 1) = 12$

d $x(4x - 5) = 2(x + 1)$

e $\dfrac{x(x - 6)}{2} = 8$

f $\dfrac{x}{16} = \dfrac{1}{3x - 8}$

3 Solve these equations using factorisation.

a $p^2 + 10p - 56 = 0$

b $q^2 + 72 = 22q$

c $2r^2 = 9(5 - r)$

4 It is given that 2 is a solution to the equation $x^2 - 15x + c = 0$, for c a constant.

a Show that $c = 26$

b Find the other solution to this equation.

5 It is given that -2 is a solution to the equation $2x^2 + (k + 1)x - k = 0$, for k a constant.

Find the other solution to this equation.

6 a By replacing x^2 with y, express $x^4 - 5x^2 + 4 = 0$ as a quadratic equation in y.

b Find the possible values of y and hence solve the equation $x^4 - 5x^2 + 4 = 0$.

7 By making a suitable substitution, solve these equations.

a $x^4 - 13x^2 + 36 = 0$

b $x - 4\sqrt{x} + 3 = 0$

c $2^{2x} - 9 \times 2^x + 8 = 0$

d $3 \times 9^x - 10 \times 3^x + 3 = 0$

> **Handy hint**
>
> For part **c**, let $y = 2^x$.

8 The diagram shows a rectangular field $ABCD$.

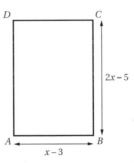

$AB = (x - 3)$ km and $BC = (2x - 5)$ km.

The area of the field is 1 km².

a Show that x satisfies the equation $2x^2 - 11x + 14 = 0$.

b Find the value of x and explain it is the only possible value.

c Hence find the perimeter of this field.

9 The diagram shows a rectangle $ABCD$.

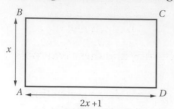

$AB = x$ cm and
$AD = (2x + 1)$ cm.

The area of this
rectangle is 15 cm^2.

Find the length of the
diagonal BD.

10 The diagram shows a rectangle $ABCD$.

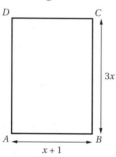

$AB = (x + 1)$ cm and $BC = 3x$ cm.

Find the area of this rectangle given that its
area and perimeter are numerically equal.

PS 11 The diagram shows the trapezium $PQRS$.

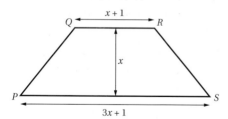

The trapezium has height x cm and
horizontal sides.

$PS = (3x + 1)$ cm and $QR = (x + 1)$ cm.

a Show that, in cm^2, an expression for the
area of this trapezium is $x(2x + 1)$.

The area of this trapezium is 28 cm^2.

b Find the area of triangle PRS.

c Given that $PQ = SR$, show that this
trapezium has perimeter $16 + 7\sqrt{2}$ cm.

12 Find the perimeter of these right-angled
triangles. All lengths are in cm.

a

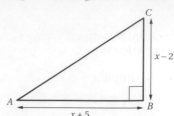

b

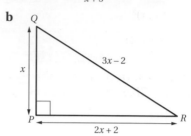

13 The diagram shows a right-angled triangle
PS ABC.

$AB = x^2 - 1$ and $AC = 2x$, where all lengths are
in cm.

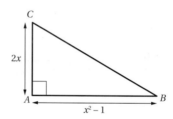

a Show that $BC = x^2 + 1$.

The perimeter of this triangle is 112 cm.

b Find the area of this triangle.

4.2 Using the quadratic formula

1 Use the quadratic formula to find the exact
solutions of these equations.

Simplify each answer as far as possible.

a $x^2 + 6x + 1 = 0$

b $x^2 + 4x - 3 = 0$

c $x^2 - 6x + 3 = 0$

d $2x^2 + x - 2 = 0$

e $3x^2 - 4x = 2$

f $2x^2 = 10x + 3$

2 Simplify these equations and then solve them using the quadratic formula.

Give answers in simplified surd form.

a $3x^2 - 2x = 2(x^2 + 3)$

b $4x^2 + 1 = (2 - x)^2$

c $2x(x - 3) = (x + 1)(x + 3)$

3 Use the quadratic formula to find the roots of these equations. Give answers to 2 decimal places.

a $x^2 - 5x + 3 = 0$

b $x(3x - 8) + 3 = 0$

c $2x(x - 3) = 17$

4 How many real roots do these quadratic equations have?

a $x^2 + 3x - 2 = 0$

b $2x^2 - 5x + 4 = 0$

c $4x^2 - 4x + 1 = 0$

> **Handy hint**
>
> Calculate the discriminant of each equation.

5 a Show that the roots of the equation $x^2 - 2kx - 1 = 0$, where k is a constant, are given by $x = k \pm \sqrt{k^2 + 1}$.

b Find, in terms of k, the sum of these two roots.

c Find the product of these two roots.

6 Find the root(s) of these quadratic equations, leaving your answer(s) in terms of the positive constant k. Simplify each answer as far as possible.

a $x^2 - kx + 2 = 0$

b $2x^2 - 3\sqrt{k}x + k = 0$

c $k^2x^2 + 2kx + 1 = 0$

d $4x^2 - 4kx + (k^2 - 9) = 0$

7 The equation $x^2 - 4x + 2p = 0$ has no real roots, where p is a real number.

Show that $p > 2$.

8 a Find the value of the smallest positive integer q for which the equation $2x^2 + qx + 7 = 0$ has two real roots.

b For this value of q find these two roots.

9 The diagram shows a shape consisting of a circle with radius r cm on top of a rectangle $ABCD$. The rectangle has height 10 cm and width equal to the diameter of the circle.

The area of the shape is 240 cm².

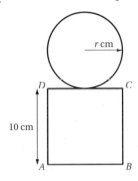

a Find the value of r, giving your answer to 1 decimal place.

b Hence show that this circle occupies just under half of this shape.

10 For a particular integer c, the equation $x^2 + 8x + c = 0$ has no real roots.

For this integer c, the equation $x^2 + 12x + 2c = 0$ has two real roots, p and q, where $p < q$.

Find the value of p and the value of q, simplifying each answer as far as possible.

11 The diagram shows a line passing through the points $A(1, 3)$ and $B(p, p)$, where $p > 3$ is a constant.

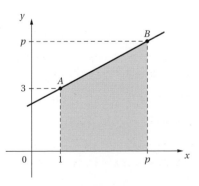

Given that $AB = 6$

a show that $p^2 - 4p - 13 = 0$

Handy hint
See Section 3.3.

b Find the area of the trapezium shaded in the diagram.

Give your answer in the form $m + n\sqrt{17}$ for integers m and n to be stated.

(PS) 12 The diagram shows a line passing through the points $A(2, 4)$, $B(p, p)$ and $C(q, 7)$, where p and q are constants, and $p > 2$.

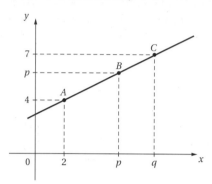

Given that $AB = 4$:

a Show that $p = 3 + \sqrt{7}$.

b Find the exact value of q.

c By finding the distances AC and BC show that $\sqrt{4 + \sqrt{7}} - \sqrt{4 - \sqrt{7}} = \sqrt{2}$.

(PS) 13 The diagram shows a right-angled triangle ABC.

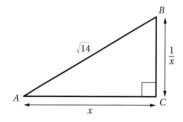

$AB = \sqrt{14}$ cm, $AC = x$ cm and $BC = \dfrac{1}{x}$ cm, where $x > 1$.

Handy hint
See Practice 4.1 Q6.

a Express $\left(2 + \sqrt{3}\right)^2$ in the form $a + b\sqrt{3}$ for integers a and b.

b Show that x satisfies the equation $x^4 - 14x^2 + 1 = 0$.

c Find the exact value of x and hence show that the perimeter of this triangle is $4 + \sqrt{14}$ cm.

4.3 Further equation solving

1 Solve these simultaneous equations.

a $y = x + 3$
$x^2 + y^2 = 5$

b $y = 2x + 1$
$x^2 + y^2 = 10$

c $y = 3x + 2$
$2x^2 + y^2 = 3$

2 The diagram shows the circle $x^2 + y^2 = 5$ and the line $y = 3 - 2x$. Points A and B are where the line and the circle intersect.

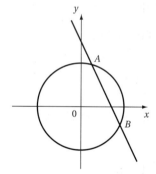

a Solve the simultaneous equations
$x^2 + y^2 = 5$
$y = 3 - 2x$

to find the coordinates of point A and the point B.

b Show that the distance AB is $\dfrac{8\sqrt{5}}{5}$ units.

3 Solve these simultaneous equations.

a $y - 2x + 1 = 0$
$y^2 - x^2 + 2x = 9$

b $x - 2y - 1 = 0$
$x^2 + 4y^2 = 5$

c $2x + 3y + 1 = 0$
$4x^2 - y^2 = 15$

4 The diagram shows the circle $x^2 + y^2 = 20$ and the line with equation $y = 2x + 10$.

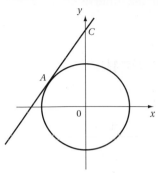

The line crosses the y-axis at point C and is a tangent to the circle at point A.

a Write down the coordinates of point C.

b Solve the simultaneous equations
$$x^2 + y^2 = 20$$
$$y = 2x + 10$$
to find the coordinates of point A.

c Find the area of triangle OAC.

5 A circle has equation $x^2 + y^2 = 10$ and a line has equation $y + 3x = 10$.

a Solve the simultaneous equations
$$x^2 + y^2 = 10$$
$$y + 3x = 10$$
to find the coordinates of any points where this line and circle intersect.

b What information about this line and circle does your answer to **a** give?

6 A circle has equation $x^2 - 2x + y^2 = 4$ and a line has equation $2y + x = 7$.

a Show that the simultaneous equations
$$x^2 - 2x + y^2 = 4$$
$$2y + x = 7$$
have no real solutions.

b What information about this line and circle does your answer to **a** give?

7 Solve these cubic equations.

a $(x + 4)(x + 2)(x - 3) = 0$

b $x(x + 3)(x - 2) = 0$

c $(3x - 2)(x + 2)(5 - x) = 0$

d $(x + 1)(x^2 - 5x + 6) = 0$

e $x^3 + 3x^2 - 4x = 0$

f $4x^3 - 3x^2 = 0$

g $3x^3 - 10x^2 - 8x = 0$

h $4x^3 + 9x = 12x^2$

8 The shape in the diagram consists of two square metal plates welded together. The square $ABCG$ has side length x metres. The smaller square $CDEF$ has side length y metres.

(PS)

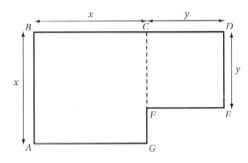

The shape has perimeter 10 m and area 5 m²

a Use this information to write down and solve a pair of simultaneous equations.

b Find the area of each plate.

9 The diagram shows a right-angled triangle ABC. $AB = x$ cm, $AC = y$, where $x > y$, and $BC = 6$ cm.

(PS)

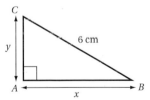

The triangle has perimeter 14 cm.

a Write down and solve a pair of simultaneous equations to find the exact length of side AB and the exact length of side AC.

b Find the area of triangle ABC.

10 The diagram shows a rectangular field *ABCD*.

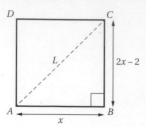

$AB = x$, $BC = (2x - 2)$ and the diagonal *AC* has length *L*, where all lengths are in kilometres.

a Show that $L^2 = 5x^2 - 8x + 4$.

A hiker knows that the route *ACDBA* is 2 km longer than the route *ABCDA*, where all sections of each journey consist of straight lines.

b Use this information to express *L* in terms of *x*.

c Find the area of the field.

5 Practice: Coordinate geometry 2

5.1 Transformations of graphs

1 The diagram shows the graph of $y = f(x)$ which crosses the y-axis at the point $(0, 7)$. Point $P(-2, 3)$ is the minimum point on this graph.

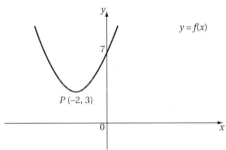

Sketch, on separate diagrams, the graphs of:

a $y = f(x) + 2$ b $y = 2f(x)$

c $y = f(2x)$ d $y = f(x - 2)$.

Indicate on each sketch the coordinates of the minimum point and the y-intercept of each graph.

2 The diagram shows the graphs F and G. The graph G is a translation of graph F by the vector $\begin{pmatrix} 3 \\ -4 \end{pmatrix}$.

> Refer to Section 5.1 if you are not sure what this means.

Under this translation, point $P(1, 4)$ on F is mapped to point P' on G, and point Q on F is mapped to point $Q'(6, -2)$ on G.

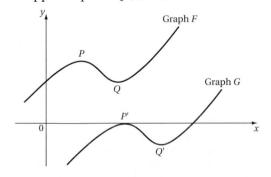

a Find the coordinates of:

   i point P' ii point Q.

b Sketch the graph of $y = \frac{3}{2} f(x)$.

   Label the images of P and Q with their coordinates.

3 The diagram shows the graph of $y = f(x)$ and the graph of $y = f(ax)$, where a is a positive constant. The graph $y = f(x)$ crosses the x-axis at the points where $x = -1$ and $x = 6$. Both graphs cross the y-axis at the point $(0, 2)$.

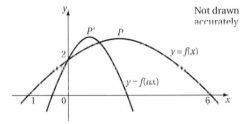

Not drawn accurately

a Describe, in terms of a, the transformation which maps the graph of $y = f(x)$ onto the graph of $y = f(ax)$.

   Under this transformation, the point $P(2, 3)$ is mapped to the point $P'\left(\frac{1}{2}, 3\right)$.

b Find the value of a.

c For this value of a, find the coordinates of the points where the graph of $y = f(ax)$ crosses the x-axis.

d Sketch, on separate diagrams, the graphs of

   i $y = f(x + 2)$ ii $y = 4f(4x)$.

   On each sketch, mark the axis-crossing points with their values and the maximum point with its coordinates.

4 The diagram shows the graph of $y = f(x)$. The graph crosses the y-axis at the point $(0, 3)$ and the x-axis at the point $(5, 0)$. The maximum point P on this graph has coordinates $(2, 5)$.

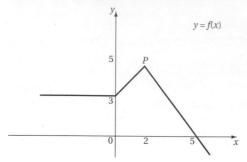

a Sketch, on separate diagrams, the graphs of:
  i $y = -f(x)$ **ii** $y = f(-x)$.

Indicate on each sketch the coordinates of the axis-crossing points and the maximum or minimum point of each graph.

b Sketch, on a new diagram, the graph of $y = -f(-x)$. Indicate the coordinates of the axis-crossing points and minimum point of this graph.

c Describe in as much detail as you can the **single** transformation which maps the graph of $y = f(x)$ to the graph of $y = -f(-x)$.

(PS) 5 The diagram shows the complete graph of $y = f(x)$. The graph crosses the y-axis at the point $(0, 5)$ and the x-axis at the points $(2, 0)$ and $(6, 0)$. Point $P(4, -2)$ is the minimum point on this graph.

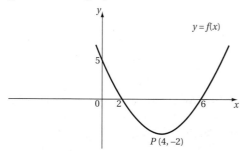

a Sketch, on separate diagrams, the graphs of:
  i $y = f(x) + 3$ **ii** $y = f\left(\frac{1}{2}x\right)$
  iii $y = -f(2x)$ **iv** $y = 2f(3x)$.

Indicate on each sketch the coordinates of any axis-crossing points and the minimum or maximum point of each graph.

Under the translation $\begin{pmatrix} -2 \\ k \end{pmatrix}$, where $k > 0$, the graph of $y = f(x)$ is mapped to a graph G which crosses the x-axis at a single point.

b State the value of k.

c Find the coordinates of the points where G crosses each axis.

5.2 Sketching curves

1 On separate diagrams sketch the curves with these equations. On each sketch, label the turning point with its coordinates.

  a $y = x^2 + 4$

  b $y = (x - 4)^2$

  c $y = (x + 3)^2 + 2$

2 By writing each quadratic expression as a completed square, sketch, on separate diagrams, the curves with these equations. On each sketch label the turning point with its coordinates.

  a $y = x^2 + 8x + 20$

  b $y = x^2 + 2x - 5$

  c $y = x^2 - 6x + 7$

  d $y = x^2 - 3x + 3$

3 By factorising each quadratic expression, sketch, on separate diagrams, the curves with these equations. On each sketch label the y-intercept with its value.

  a $y = x^2 - 5x + 4$

  b $y = 2x^2 + 7x + 5$

  c $y = -x^2 + 7x - 10$

  d $y = 12 + 5x - 2x^2$

4 On separate diagrams sketch the curves with these equations. On each sketch label the y-intercept with its value.

  a $y = (x + 3)(x - 2)(x - 1)$

  b $y = (2x - 1)(x + 3)(x + 1)$

  c $y = (x - 1)^2(x - 3)$

  d $y = (x + 2)(x^2 - 4)$

5 By factorising each cubic expression as far as possible, sketch, on separate diagrams, the curves with these equations.

a $y = x^3 + 4x^2 + 3x$

b $y = x^3 + 2x^2 - 8x$

c $y = x^3 + 4x^2 + 4x$

d $y = 4x^3 - 12x^2 + 9x$

6 The diagram shows the curve with equation $y = x^2 + bx + c$, where b and c are constants.

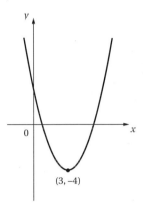

The turning point of this graph has coordinates $(3, -4)$

a Find the value of b and the value of c

b Hence express the equation of this curve in the form $y = (x - r)(x - s)$ for constants r and s.

7 a Expand and then simplify $(x - 1)(x^2 + x - 12)$

b Hence sketch the curve with equation $y = x^3 - 13x + 12$

8 The diagrams show the graphs of four cubic curves together with their roots and y-intercepts.

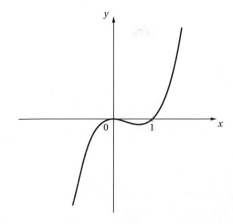

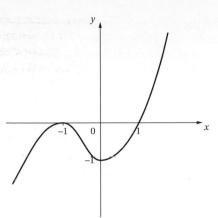

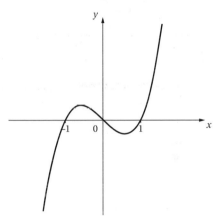

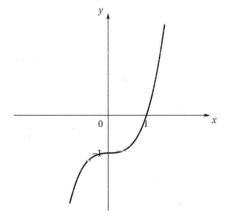

The equations of three of these curves are

A: $y = x^3 - 1$ B: $y = x^3 - x$ C: $y = x^3 - x^2$

a Match each equation A, B and C to its corresponding diagram.

b Suggest an equation for the curve shown on the unmatched diagram.

9 Each of the curves shown in these diagrams has an equation of the form $y = x^2 + bx + c$ where b and c are constants. P is the turning point of each curve.

For each curve find the value of b, c and any other unknown values shown on the diagram.

> **Handy hint**
> The roots of a quadratic curve are equally spaced either side of its line of symmetry.

a

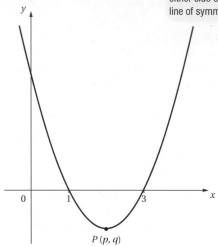

$P(p, q)$

b

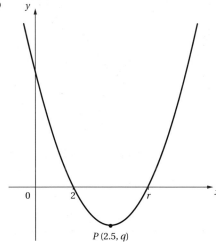

$P(2.5, q)$

c

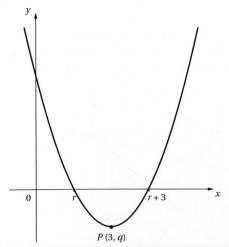

$P(3, q)$

d

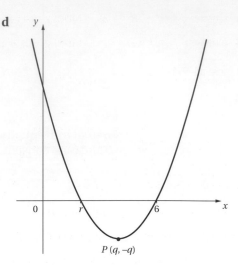

$P(q, -q)$

10 a Sketch the graph with equation $y = x^2 - 8x + 15$, labelling its roots and turning point with their coordinates.

b Hence, or otherwise, state the value of the constant k such that the graph with equation $y = x^2 - 6x + 8 + k$ has exactly one real root.

c Given that the graph with equation $y = (x - a)(x - b) + k$ has exactly one real root, where $a < b$ and k are constants, show that $k = \dfrac{(b-a)^2}{4}$.

5.3 Intersection points of graphs

1 The diagram shows the curve $y = x^2 - 6x + 13$ and the line $y = 2x + 1$. The curve and line intersect at points A and B.

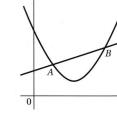

a Find the coordinates of A and B.

b Show that the distance $AB = 4\sqrt{5}$ units.

2 Find the coordinates of the points of intersection of these curves and lines. Give answers in simplified surd form where appropriate.

a Curve: $y = x^2 - 3x - 4$ Line: $y = 4 - x$

b Curve: $y = x^2 + 5x + 7$ Line: $y = 2x + 7$

c Curve: $y = x^2 + 6x + 3$ Line: $y = 4x + 4$

d Curve: $y = \frac{1}{6}x^2$ Line: $y - x + 1 = 0$

3 The diagram shows the curve
$y = -x^2 + 7x - 6$
and the line $y = 4$.
The curve crosses
the x-axis at points
A and B. The curve
and line intersect
at points C and D.

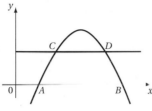

a Find the coordinates of A and B.

b Find the coordinates of points C and D.

c Hence find the area of the trapezium $ABCD$.

4 The diagram shows the curve
$y = 2x^2 - 4x + 5$ and the line $y = 8x - 13$.
The curve crosses the y-axis at point A.
The line crosses the y-axis at point B.

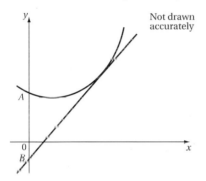

Not drawn accurately

a Write down the coordinates of A and B.

b Show that this line and curve intersect at a
single point, P, and give the coordinates of P.

c Show that triangle ABP has area 27 square
units.

5 A curve has equation $y = x^2 - 4x + 7$ and a
line has equation $y = 1 - 2x$.

a Use the method of substitution to show
that the simultaneous equations
$y = x^2 - 4x + 7$ and $y = 1 - 2x$ have no
real solutions.

b What information about this curve and line
does the result of part **a** give?

c Express $x^2 - 4x + 7$ in the form
$(x - p)^2 + q$, where p and q are constants.

d Sketch, on the same diagram, the graph of
$y = x^2 - 4x + 7$ and the line $y = 1 - 2x$.

6 a Sketch the graph of $y = x^2 + 6x + 13$.
Label the vertex of your sketch with its
coordinates.

b Find the range of values of k for which the
horizontal line $y = k$ does not intersect the
curve $y = x^2 + 6x + 13$. Give your answer as
an inequality.

7 Find the coordinates of the points of
intersection of:

Handy hint

See Section 4.3.

a the curve with equation
$y = x^3 + 3x^2 + 1$ and the
line with equation $y + 2x = 1$.

b the curve with equation
$y = 2x^3 - 4x^2 - 3x + 1$ and the curve with
equation $y = 5x^2 - 12x + 1$.

8 The diagram shows the graph of
(PS) $y = x^2 - 8x + 21$ and $y = -x^2 + 6x + 9$.
The curves intersect at points A and B.

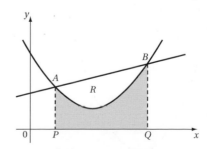

Not drawn accurately

a Find the coordinates of A and B.

The line L is the perpendicular bisector of AB.

b Find, to 1 decimal place, the coordinates
of the points where L intersects the curve
$y = -x^2 + 6x + 9$.

9 The diagram shows the graph of
(PS) $y = x^2 - 6x + 13$ and the line $y = 2x + 6$.
The line intersects the curve at points A and B.

Points P and Q on the x-axis are such that the lines PA and QB are vertical. The shaded region bounded by the curve, the x-axis and the lines AP and BQ is 48 square units.

Find the area of the region R between the curve and the line AB.

PS 10 a Find the coordinates of the points where the curve with equation $y = (x - 1)^2 (x + 1)$ intersects the curve with equation $y = 3x^2 + 4x + 1$

b Sketch, on a single diagram, these two curves, showing clearly these intersection points.

See Section 5.2.

6 Practice: Trigonometry

6.1 Trigonometry and triangles

Unless told otherwise, give final answers to 3 significant figures.

1 In these triangles, all lengths are in centimetres. Use the sine rule to find the length of the side indicated with a lowercase letter.

a

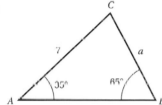

b

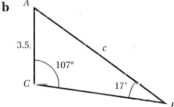

c

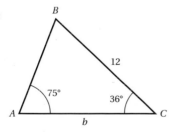

d

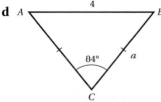

> **Handy hint**
>
> The dashes on sides AC and BC mean $AC = BC$.

2 The diagram shows triangle ABC. $AC = 6$ cm, $BC = 10$ cm and angle $BAC = 74°$.

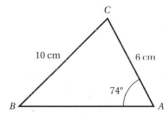

a Show that angle $CBA = 35.2°$ to 3 significant figures.

b Use the sine rule to find the length AB.

3 The diagram shows triangle ABC. $AB = x$, $AC = 2x$ and angle $ACB = 30°$.

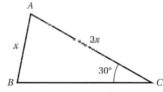

a Use the sine rule to show that $\sin \hat{B} = 1$.

b Hence show that triangle ABC is right-angled.

c Express the length of the side BC in terms of x, simplifying your answer as far as possible.

4 Rearrange these cosine rules to make the required term the subject.

a $c^2 = a^2 + b^2 - 2ab \cos \hat{C}$ for $\cos \hat{C}$

b $b^2 = a^2 + c^2 - 2ac \cos \hat{B}$ for $\hat{B}$

5 In these triangles, all lengths are in centimetres. Use the cosine rule to find the length of the side indicated with a lowercase letter.

a

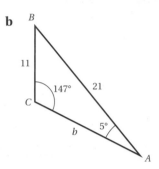

b

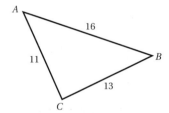

6 In any triangle, the largest angle is opposite the longest side.

a Use the cosine rule to find the largest angle in this triangle.

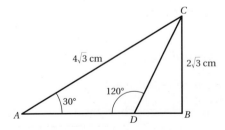

b Use any appropriate rules to find the other two angles of this triangle.

7 The diagram shows triangle ABC where $AC = 4\sqrt{3}$ cm , $BC = 2\sqrt{3}$ cm and angle $CAB = 30°$.

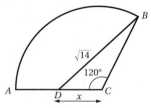

Point D on AB is such that angle $ADC = 120°$.

a Show that $CD = 4$ cm.

b Find angle DBC.

c Find the exact perimeter of triangle ABC.

8 The diagram shows the triangle ABC, where $AC = 2\sqrt{3}x$ cm , $AB = 6x$ cm and angle $BAC = 30°$.

(PS)

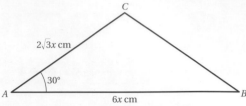

a Show that triangle ABC is isosceles.

b Hence, or otherwise, show that the area of triangle ABC is $\sqrt{k}x^2$ cm², stating the value of the integer k.

9 The diagram shows the triangle ABC, where $AB = 7$ cm, $AC = 12$ cm and angle $BAC = 25°$. Point D on AC is such that $AD = 4$ cm and angle $DBC = 50°$.

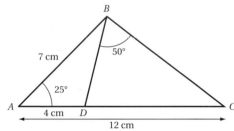

a Show that $BD = 3.77$ cm (to 3 significant figures).

b Find angle C.

c Hence, or otherwise, find angle DBA.

10 The diagram shows the sector of a circle with centre C. Points A and B lie on this circle. $DB = \sqrt{14}$ cm, where D is the mid-point of AC. $DC = x$ cm and angle $DCB = 120°$.

(PS)

a Show that $x = \sqrt{2}$ cm.

b Find the perimeter of the curved shape ADB.

PS 11 The diagram shows the triangle *ABC* where
$AB = k$ cm, $BC = (k + 1)$ cm and
$AC = (2k - 1)$ cm. Angle $ABC = 120°$.

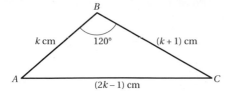

a Show that $k = 7$.

b Find the other two angles in this triangle.

Give each answer to the nearest degree.

PS 12 The diagram shows a quadrilateral *ABCD*. All
lengths are in centimetres. $AB = 10$, $BC = 18$,
$CD = 12$ and $DA = 14$.

Angle $ABD = 27°$.

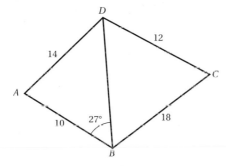

a Show that angle $ADB = 18.9°$ (to 3
significant figures).

b Find angle *DCB*.

c Explain briefly why the points *A*, *B*, *C* and *D*
cannot all lie on a common circle.

d Show that the diagonal *AC* has length
16 cm to the nearest centimetre.

6.2 The area of any triangle

Unless told otherwise, give final answers to 3
significant figures.

1 Find the area of each of these triangles.

a

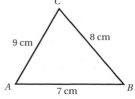

b

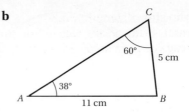

c

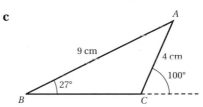

2 Find the area of this
isosceles triangle.

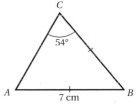

3 The diagram shows triangle *PQR*.
$PQ = 10$, $PR = 30$ and angle $PQR = 150°$.
All lengths are in centimetres.

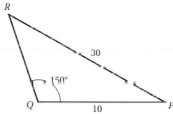

a Use the sine rule to show that $\sin \hat{R} = \frac{1}{6}$.

b Hence find angle *QRP*.

c Show that triangle *PQR* has area 52.3 cm²
to 3 significant figures.

4 Find the area of an equilateral triangle which
has perimeter 15 cm.

5 In triangle *ABC*,
$AB = 7$ cm
$BC = 8$ cm
$AC = 9$ cm.

a Use the cosine rule to show that angle
$ACB = 48.2°$ (to 3 significant figures).

b Hence find the area of this triangle.

6 The diagram shows
a circle, radius 8 cm
and centre at
point C.
Points A and B
on the circle are
such that angle
$ACB = 45°$.

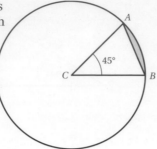

a Find the area
of triangle ABC.

b Show that the shaded segment between
this circle and the line AB has area
2.5 cm^2 (to 1 decimal place).

7 The triangles ABC and PQR shown in the
diagram have equal areas.

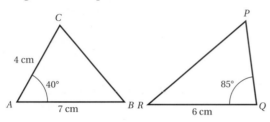

a Find the area of triangle ABC.

b Hence find the length of the side QP.

c Show that the perimeter of triangle PQR is
15.5 cm (to 3 significant figures).

8 The diagram shows
the triangle PQR where
$PQ = x$, $PR = 2x$ and
angle $QPR = 30°$.
All lengths are in
centimetres.

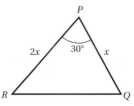

It is given that the area of this triangle is
18 cm^2 .

a Find the length of the side PQ.

b Hence show that the base RQ of this triangle
has length 7.44 cm (to 3 significant figures).

c Find the length of the shortest line from P
to the side RQ.

9 The diagram shows the
design of a badge
in the shape of a
sector ABC of a
circle. The circle
has centre A and
radius 10 cm.
Angle BAC is 85°.

(PS)

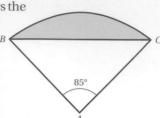

Show that the area of the shaded region is
24.4 cm^2 to 1 decimal place.

10 A metal plate is formed by welding triangle
ABC to a sector BCD of a circle. The circle has
centre C and radius 7 cm. Angle BAC is 25°
and angle BCD is 55°.

(PS)

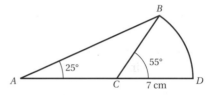

a Find the area of the plate.

b Find the perimeter of the plate.

11 The diagram shows a rectangle $ABCD$,
where $AB = 4$ cm. Point E on BC is such that
$BE = 3$ cm and point F on CD is such that
$EF = \sqrt{2}$ cm and $AF = \sqrt{17}$ cm.

(PS)

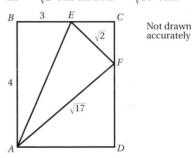

Not drawn
accurately

a Show that angle $AEF = 45°$.

b Find the area of triangle AEF.

c Using calculator accuracy, show that
$EC = \frac{1}{5}$.

d Hence find the area of triangle AFD.

6.3 Solving a trigonometric equation

Where appropriate, give answers to 1 decimal place.

1 The diagram shows the graph of $y = \sin x$ for $0° \leqslant x \leqslant 360°$.

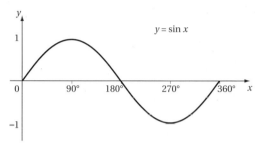

a Use this graph to write down the solutions to the equation $\sin x = 0$ for $0° \leqslant x \leqslant 360°$.

b Use your calculator to find one solution to the equation $\sin x = 0.6$.

c Use this graph to find another solution to the equation $\sin x = 0.6$ for $0° \leqslant x \leqslant 360°$.

2 The diagram shows the graph of $y = \cos x$ for $0° \leqslant x \leqslant 360°$.

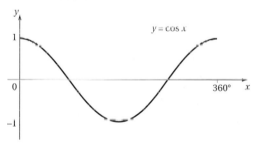

a Write down the solution of the equation $\cos x = -1$ for $0° \leqslant x \leqslant 360°$.

b Solve the equation $\cos x = -0.7$ for $0° \leqslant x \leqslant 360°$.

3 By using a calculator and the properties of the sine graph, solve these equations.

a $\sin x = 1$ for $0° \leqslant x \leqslant 360°$

b $\sin x = 0.25$ for $0° \leqslant x \leqslant 360°$

c $\sin x = 0.5$ for $0° \leqslant x \leqslant 540°$

4 By using a calculator and the properties of the cosine graph, solve these equations.

a $\cos x = 1$ for $0° \leqslant x \leqslant 360°$

b $\cos x = \frac{2}{3}$ for $0° \leqslant x \leqslant 360°$

c $\cos x = -\frac{1}{\sqrt{2}}$ for $0° \leqslant x \leqslant 720°$

5 The diagram shows the graph of $y = \sin x$ for $0° \leqslant x \leqslant 360°$.

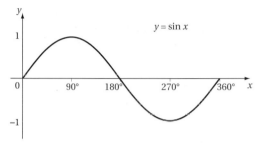

a Use your calculator to verify that $\sin 45° = \cos 45°$.

b On a copy of this diagram, sketch the graph of $y = \cos x$ for $0° \leqslant x \leqslant 360°$.

c Use the result of part **a** to write down a solution to the equation $\sin x = \cos x$. Illustrate this solution on your sketch.

d Use your sketch to find the other solution to the equation $\sin x = \cos x$ for $0° \leqslant x \leqslant 360°$. Verify your answer using a calculator.

6 The diagram shows triangle ABC where
(PS) $AC = 8$ cm, $BC = 7$ cm and angle $CAB = 30°$. The triangle has been drawn so that angle ABC is acute.

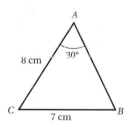

a Use the sine rule to show that $\sin \hat{B} = \frac{4}{7}$.

b Hence find the acute angle B.

c Find an obtuse angle which satisfies the equation $\sin x = \frac{4}{7}$.

d Using your answer to part **c**, sketch a triangle PQR, not congruent to triangle ABC, for which $PR = 8$ cm, $QR = 7$ cm and angle $RPQ = 30°$.

7 Solve these equations for $0° \leqslant x \leqslant 360°$.

   a $2 \sin x = 1$ **b** $3(1 + \cos x) = 4$

   c $3 + \sqrt{2} \sin x = 4$ **d** $\dfrac{3 - 4 \cos x}{2} = 1$

8 **a** Show, with the aid of a sketch, that
$\sin(x + 90°) = \cos x$ for all values of x.

   b Hence solve the equation $\sin(x + 90°) = 0.2$
for $0° \leqslant x \leqslant 360°$.

   c Solve, using a similar method, the equation
$\cos(x + 90°) = -0.75$ for $0° \leqslant x \leqslant 360°$.

9 **a** Use your calculator to find
a solution of the equation
$\sin x = -0.5$.

   Handy hint
   For part **a**, illustrate
   this solution on a
   sketch of $y = \sin x$.

   b Use the graph of $y = \sin x$
to find the smallest positive solution to
the equation $\sin x = -0.5$.

   c Find the two smallest positive solutions to
the equation $\sin x = -\dfrac{4}{5}$.

10 **a** Show, with the aid of a sketch, why the
equation $\sin x = \dfrac{5}{4}$ has no solutions.

   b Solve, where possible, these equations for
$0° \leqslant x \leqslant 360°$.

     i $2 \cos x = 3$ **ii** $4 \sin x + 3 = -1$

    iii $\dfrac{\sin x}{3} = -0.334$ **iv** $\dfrac{3}{\cos x} - 1 = 3$

11 **a** Describe the transformation which maps
the graph of $y = \sin x$ to
the graph of $y = \sin ax$,
where a is a positive
constant.

   Handy hint
   See Section 5.1.

   b Solve the equation $\sin x = 0.4$ for
$0° \leqslant x \leqslant 360°$.

   c Hence, or otherwise, solve these equations.

     i $\sin 2x = 0.4$ for $0° \leqslant x \leqslant 180°$

     ii $\sin 4x = 0.4$ for $0° \leqslant x \leqslant 90°$

    iii $\sin \dfrac{3x}{2} = 0.4$ for $0° \leqslant x \leqslant 240°$

12 **a** Solve the equation
$\cos x = -\dfrac{3}{8}$ for $0° \leqslant x \leqslant 360°$.

   b Hence, or otherwise, solve these equations.

     i $\cos 3x = -\dfrac{3}{8}$ for $0° \leqslant x \leqslant 120°$

     ii $8 \cos 2x + 3 = 0$ for $0° \leqslant x \leqslant 180°$

13 Solve these equations for $-360° \leqslant x \leqslant 360°$.

   a $\cos x = 0.28$ **b** $\sin x = 0.65$

   c $\sin x = -0.7$ **d** $5 \cos x + 3 = 1$

7 Practice: Vectors

7.1 The magnitude and direction of a vector

All directions $0° \le \theta < 360°$ are measured against the positive horizontal.

Where appropriate, give magnitudes in simplified surd form unless told otherwise.

1 Find the magnitude and direction of these vectors. When not exact, give angles to 1 decimal place.

a $\begin{pmatrix} 3 \\ 3 \end{pmatrix}$ **b** $\begin{pmatrix} -5 \\ 12 \end{pmatrix}$ **c** $\begin{pmatrix} 6 \\ -2 \end{pmatrix}$ **d** $\begin{pmatrix} -\sqrt{6} \\ -\sqrt{2} \end{pmatrix}$

2 Given that $\mathbf{p} = \begin{pmatrix} 3 \\ 5 \end{pmatrix}$ and $\mathbf{q} = \begin{pmatrix} 5 \\ 3 \end{pmatrix}$ find the magnitude and direction of these vectors.

a $\mathbf{p} + \mathbf{q}$ **b** $\mathbf{p} - \mathbf{q}$ **c** $3\mathbf{p} - 5\mathbf{q}$ **d** $\frac{1}{15}(5\mathbf{p} - 3\mathbf{q})$

3 Express these vectors in component form.

a The vector $\mathbf{p}$, magnitude 6, direction 30°.

b The vector $\mathbf{q}$, magnitude 4, direction 120°.

c The vector $\mathbf{r}$, magnitude 5, direction 270°.

d The vector $\mathbf{s}$, magnitude $\sqrt{2}$, direction 315°.

4 The vectors $\mathbf{p} = \begin{pmatrix} 2 \\ \sqrt{12} \end{pmatrix}$ and $\mathbf{q} = \begin{pmatrix} -\sqrt{3} \\ 1 \end{pmatrix}$.

a By finding their directions, show that $\mathbf{p}$ and $\mathbf{q}$ are perpendicular.

b Find $|\overrightarrow{PQ}|$, where $\overrightarrow{OP} = \mathbf{p}$ and $\overrightarrow{OQ} = \mathbf{q}$.

> **Handy hint**
> Two vectors are **perpendicular** if the angle between the lines representing them is 90°.

5 Given vectors $\mathbf{p} = 5\mathbf{i} - 3\mathbf{j}$, $\mathbf{q} = 2\mathbf{i} + 2\mathbf{j}$:

a Find the magnitude and direction of these vectors. Give answers to 1 decimal place.

i $\mathbf{p} + 2\mathbf{q}$ **ii** $3\mathbf{p} - 2\mathbf{q}$ **iii** $4\mathbf{q} - 3\mathbf{p}$

b Find the vector $\mathbf{r}$ such that $2\mathbf{p} + 7\mathbf{q} + 4\mathbf{r} = \mathbf{0}$, where $\mathbf{0}$ is the zero vector. Give your answer in $\mathbf{i} - \mathbf{j}$ form.

6 a Given that $|\mathbf{p}| = 13$, where $\mathbf{p} = 5\mathbf{i} + a\mathbf{j}$, find the value of the positive constant a.

b Given that the direction of $\mathbf{q} = b\mathbf{i} + \sqrt{12}\mathbf{j}$ is 60°, find the value of b.

c Find the direction of $\mathbf{r} = c\mathbf{i} + d\mathbf{j}$, for c and d positive constants, given that $|\mathbf{r}| = 2d$.

7 Try this question **without** using a calculator.

(PS) The vector $\mathbf{c} = \begin{pmatrix} 1 \\ 3 \end{pmatrix}$ and the vector $\mathbf{d}$ has magnitude $\sqrt{6}$, direction 45°.

> **Handy hint**
> If needed, revise your trig ratios, for example, $\tan 60° = \sqrt{3}$ them is 90°.

a Find the vector $\mathbf{d}$. Give components in surd form.

b Find the direction of $\mathbf{c} + \mathbf{d}$.

c Find the magnitude of $\mathbf{c} + \mathbf{d}$.

8 The diagram shows triangle UVW.

(PS) The line WU passes through the origin O and makes an angle of 45° against the positive x-axis.

$OU = 8$ cm and $OW = 3$ cm.

> **Handy hint**
> For part **c** use $3 + \sqrt{3} = \sqrt{3}(\sqrt{3} + 1)$.

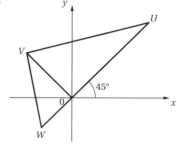

The lines OV and WU are perpendicular and the area of triangle UVW is 22 cm².

a Find these vectors. Give answers in $\mathbf{i} - \mathbf{j}$ form, leaving components as surds.

i $\overrightarrow{OU}$ **ii** $\overrightarrow{OW}$ **iii** $\overrightarrow{OV}$

b Find the exact perimeter of triangle UVW.

7.2 Position vectors

1 **a** Write down the position vectors of the points with these coordinates.

i $A(3, 2)$ **ii** $B(4, -3)$ **iii** $C\left(\sqrt{3} - 3, \sqrt{3} + 3\right)$

b Which of these three position vectors has the greatest magnitude?

c On a single diagram, sketch the position vectors of points A, B and C.

2 For each of these pairs of points find:

i $\overrightarrow{AB}$ **ii** $\left|\overrightarrow{AB}\right|$.

Give answers in simplified surd form where appropriate.

a $A(7, 2)$, $B(-3, 4)$

b $A(-3, -6)$, $B(5, -2)$

c $A\left(\sqrt{8}, -1\right)$, $B\left(\sqrt{18}, 3\right)$

3 Points A, B and C have coordinates $A(-2, 3)$, $B(1, 4)$ and $C(7, 6)$.

a Show that A, B and C are collinear:

i by finding the vectors $\overrightarrow{AB}$ and $\overrightarrow{BC}$

> *Handy hint*
> Collinear means A, B and C lie on the same straight line.

ii by finding the equation of the line which passes through points A and B.

b Comment on which of the methods used in part **a** you found easier to use.

4 **a** Given that $\left|\overrightarrow{AB}\right| = 5$ where $A(2, 7)$ and $B(k, 4)$, find the possible values of k.

b Show, on a single diagram, point A and the two possible positions of point B

c Which of these three points is furthest from the origin?

5 The diagram shows points A, B and C.

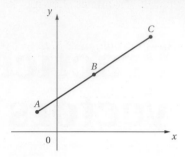

Relative to the origin O, point A has coordinates $(-1, 1)$ and point B has coordinates $(2, 3)$.

a Find the magnitude and direction of $\overrightarrow{OA}$.

Point C is such that $\overrightarrow{AC} = 2 \times \overrightarrow{AB}$.

b Find the magnitude and direction of $\overrightarrow{OC}$.

c Show that triangle AOC has area 5 square units.

6 (PS) The diagram shows triangle ABC, where, relative to the origin O, the position vectors of points A and B are $-3\mathbf{i} + \mathbf{j}$ and $\mathbf{i} + 7\mathbf{j}$, respectively.

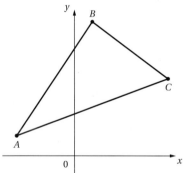

Given that $\overrightarrow{AC} = 8\mathbf{i} + k\mathbf{j}$ and $\overrightarrow{BC} = 4\mathbf{i} - 2k\mathbf{j}$.

a Show that $k = 2$.

b Find the exact lengths of the three sides of the triangle ABC.

Give each answer in the form $\sqrt{k}$, where k is an integer.

c Show that $\cos \theta = \dfrac{\sqrt{26}}{26}$, where $\theta =$ angle ABC.

d Using calculator accuracy, find the area of triangle ABC.

7 Relative to the origin O, the points A, B, C and D form a parallelogram, as shown in the diagram.

Also shown is the mid-point E of BD.

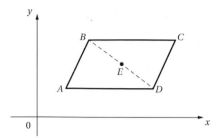

a Explain why $\overrightarrow{AB} = \overrightarrow{DC}$.

b Starting with the equation $\overrightarrow{AE} = \overrightarrow{AB} + \overrightarrow{BE}$ show that $\overrightarrow{AE} = \frac{1}{2}(\overrightarrow{DC} + \overrightarrow{BC})$.

c Hence show that $\overrightarrow{AE} = \frac{1}{2}\overrightarrow{AC}$ and explain what this result tells you about the diagonals of a parallelogram.

8 The diagram shows the points A, B and C.

Relative to the origin O, the position vectors of A and C are $2\mathbf{i} + 8\mathbf{j}$ and $12\mathbf{i} + 3\mathbf{j}$, respectively.

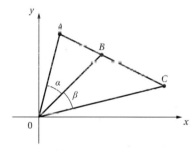

Point B on AC is such that $\overrightarrow{AB} = \frac{2}{5}\overrightarrow{AC}$.

Angle $AOB = \alpha$ and angle $BOC = \beta$.

a Show that $\overrightarrow{AB} = 4\mathbf{i} - 2\mathbf{j}$.

b Find the coordinates of point B.

c Prove that $\alpha = \beta$.

9 The diagram shows a trapezium $OABC$, where O is the origin, and sides AB and OC are parallel.

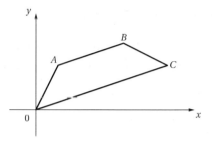

The position vectors of A and C are $\begin{pmatrix} 1 \\ 2 \end{pmatrix}$ and $\begin{pmatrix} 6 \\ 2 \end{pmatrix}$ respectively, and the y-coordinate of B is greater than 2.

a Show that the position vector of B can be written in the form

$\begin{pmatrix} 6k + 1 \\ 2k + 2 \end{pmatrix}$ for k a positive constant.

The lengths of the diagonals of this trapezium are equal.

b Find the coordinates of point B.

c Find the interior angle ABC.

8 Practice: Differentiation

8.1 Estimating the gradient of a curve

1 The diagram shows part of the curve $y = x^2$ which passes through the points $A(1, 1)$, $P(2, 4)$, and $B(3, 9)$. The tangent to the curve at P has been drawn.

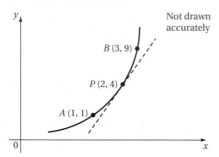

Not drawn accurately

a Find the gradient of PB.

b Find the gradient of AP.

The gradient of the tangent to the curve at P is m, where m is an integer.

c Use your answers to **a** and **b** to write down the value of m.

2 The diagram shows part of the curve $y = 4x^2 + 3$ which passes through the point $P(1, 7)$.

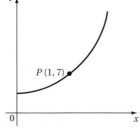

a Use each of these points to find an estimate for the gradient of the curve at P. In each case, state whether the answer gives an under- or over-estimate for this gradient.

  i $Q(0.5, 4)$ **ii** $R(1.5, 12)$

  iii The point S where $x = \sqrt{2}$, giving your answer to 3 significant figures.

b State, with a reason, which of the answers found in part **a** gives the best estimate for the gradient of the curve at P.

3 The diagram shows the curve $y = x^2 - 4x + 7$ which passes through the points $P(3, 4)$, $Q(4, 7)$ and R.

Point R is the minimum point on this curve.

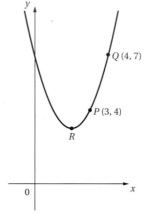

a Write down the gradient of the curve at point R.

b By expressing $x^2 - 4x + 7$ in the form $(x - a)^2 + b$, for constants a and b, find the coordinates of point R.

The gradient of the curve at point P is m, where m is an integer.

c By finding the gradients of the chords PQ and RP, show that $m = 2$.

d Write down the gradient of this curve at the point where $x = 1$.

> *Handy hint*
> Use the symmetry of the curve.

4 By calculating Grad_{PQ}, where Q is a point on the given curve, obtain an estimate for the gradient of each curve at point P.

a Curve: $y = 2x^2 - x + 4$, point $P(2, 10)$, point $Q(2.5, 14)$.

b Curve: $y = 3x - 4x^2$, point $P(2, -10)$, point $Q(1.75, -7)$.

c Curve: $y = 2x^2 + 1$, point $P(-2, 9)$, point Q with x-coordinate -1.5.

d Curve: $y = \sqrt{2x^2 + 7}$, $x \geqslant 0$ point P has x-coordinate 1, point Q has y-coordinate 5.

(PS) 5 The diagram shows part of the curve $y = \frac{1}{4}x^2 + 5$ which passes through the point P with x-coordinate 3.

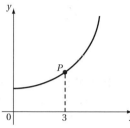

By choosing a suitable pair of points A and B on this curve either side of P, show that the gradient of the curve at P is 1.5, correct to 1 decimal place.

(PS) 6 The diagram shows part of the curve $y = 2 - \frac{3}{x}$ which passes through the point P with x coordinate 2.

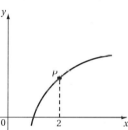

By choosing a suitable pair of points A and B on this curve either side of P, determine the gradient of the curve at point P correct to 2 decimal places.

(PS) 7 The diagram shows part of the curve $y = 2x^2$. The line $y = 5x - 3$ intersects this curve at points P and Q. The dotted line is the tangent to the curve at P. Point 0 is the origin.

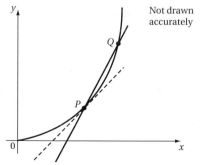

Not drawn accurately

The equation of the tangent to the curve at P is $y = mx + c$, where m is an integer and $c < 0$.

a By comparing the tangent at P with the line PQ, explain why $m < 5$.

b Use algebra to find the coordinates of P and Q.

The tangent at P intersects the line OQ at a point which has a positive x-coordinate.

c Explain why $m > 3$.

d Write down the value of m and hence find the equation of the tangent to the curve at P.

8.2 The rules of differentiation

1 Use differentiation from first principles to show that:

a if $y = 3x^2$ then $\dfrac{dy}{dx} = 6x$

b if $y = x^2 - 5x$ then $\dfrac{dy}{dx} = 2x - 5$.

2 It is given that $(x + h)^3 = x^3 + 3x^2h + 3xh^2 + h^3$.

a Find a similar expression for $(x + h)^4$.

b Use differentiation from first principles to show that if $y = x^4$ then $\dfrac{dy}{dx} = 4x^3$.

For questions 3 to 10, use the $(n - 1)$ rule to differentiate each equation.

3 Find an expression for $\dfrac{dy}{dx}$ for the curves with these equations. Factorise each answer as far as possible.

a $y = x^2 + 4x + 1$ **b** $y = (2x + 1)(2x + 3)$

c $y = x^3 - 6x^2 + 5$ **d** $y = x^3(x + 4)$

e $y = x(x + 3)(x - 5)$ **f** $y = x(x^2 + 1)^2$

4 Find the gradient of the curves with these equations at the points indicated.

a $y = x^3 + 2x^2 + 1$ at the point P where $x = 1$

b $y = \frac{1}{2}x^4 - 3x^2 - 4x$ at the point Q where $x = -2$

c $y = (x - 3)(x^2 + 9)$ at the point R where $y = 0$

d $y = \frac{1}{8}x^3 + x - 1$ at the point S where this curve intersects the line $y = x$

5 A curve has equation $y = 2x^2 + 5x - 3$.

Find the gradient of this curve at each of its roots.

6 The curve C has equation $y = 3x^2 - 2x - 1$.

   a Find the gradient of C at its y-intercept.

At point P on this curve, the gradient of C is 10.

   b Show that the coordinates of P are (2, 7).

   c Find the coordinates of the point Q on this curve where the gradient of C is -5.

(PS) 7 The curve C with equation $y = \frac{1}{3}x(x^2 - 4)$, where $x > 0$.

The gradient of C at points P and Q is $\frac{8}{3}$ and $\frac{71}{3}$, respectively.

   a Find the coordinates of P and Q.

   b Hence find the exact coordinates of the point R on C where the tangent to the curve is parallel to the chord PQ.

(PS) 8 The diagram shows the graph of $y = x^3 + 3x^2 + 2$ which passes through the points $P(1, 6)$ and Q.

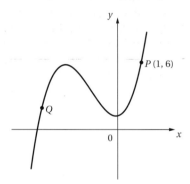

The tangents to this curve at P and Q are parallel.

Find the coordinates of point Q.

(PS) 9 The diagram shows the graph of $y = kx^3 + \frac{1}{3}x^2 + 1$, where k is a constant.

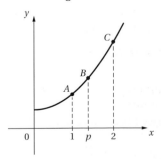

This curve passes through the points A, B and C, with x-coordinates 1, p and 2, respectively, where $1 < p < 2$.

The gradient of this curve at point A is $\frac{17}{3}$.

   a Show that $k = \frac{5}{3}$.

   b Hence find the gradient of this curve at point C.

David uses differentiation to calculate the gradient of this curve at point B.

His answer is $\frac{43}{2}$.

   c Use this answer to explain how David must have made a mistake.

David correctly recalculates this gradient. His answer this time is 8.

   d Find the coordinates of point B.

10 By expressing these equations in an appropriate form, find $\frac{dy}{dx}$.

   a $y = \dfrac{x^4 + x}{x}$ **b** $y = \dfrac{2x^5 - 3x^3 + x^2}{x^2}$

   c $y = \dfrac{4x^3 - 9x^2}{2x}$ **d** $y = \dfrac{x^2 - 4}{x + 2}$

   e $y = \dfrac{2x^2 + x - 1}{3x + 3}$ **f** $y = \dfrac{x^4 - 1}{x^2 + 1}$

11 **a** Show that $\dfrac{1}{x + h} - \dfrac{1}{x} = -\dfrac{h}{(x + h)x}$.

   b Hence, by using differentiation from first principles to show that if $y = \dfrac{1}{x}$ then $\dfrac{dy}{dx} = -\dfrac{1}{x^2}$.

   c Verify the result of part **b** by using the $(n - 1)$ rule for differentiation.

12 **a** Simplify $\left(\sqrt{x + h} - \sqrt{x}\right)\left(\sqrt{x + h} + \sqrt{x}\right)$

   b Use differentiation from first principles to show that if $y = \sqrt{x}$ then $\dfrac{dy}{dx} = \dfrac{1}{2\sqrt{x}}$.

   c Verify the result of part **b** by using the $(n - 1)$ rule for differentiation.

8.3 Applying differentiation

1 The diagram shows the curve $y = 2x^2 - 5x + 6$. The tangent T to this curve at point $P(2, 4)$ has also been drawn.

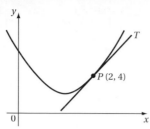

a Show that the gradient of T is 3.

b Hence find an equation for T.

2 Find the equation of the tangent to these curves at the given point P. Give each answer in the form $y = mx + c$ where m and c are constants.

a $y = 3x^2 + 2x + 5$ at the point $P(-1, 6)$

b $y = (2x - 1)(x - 1)$ at the point $P(3, 10)$

c $y = x^3 - 2x^2 + 3x - 1$ at the point $P(1, 1)$

d $y = 4 + 3x - 4x^2$ at the point where $x = 1$

3 Find the equation of the tangent to the curve $y = (x^2 + 2)(x - 3)$ at the point where this curve crosses the x-axis.

4 A curve has equation $y = 4x^2 - 6x + 1$.

a Find an equation for the tangent to this curve:

  i at the point P where $x = \dfrac{3}{2}$

  ii at the point $Q(a, 11)$, where $a > 0$.

Give each answer in the form $y = mx + c$, for constants m and c.

b Find the coordinates of the point where these two tangents intersect.

5 The diagram shows the curve $y = x^2 - 4x + 7$ which passes through the point $P(3, 4)$. The normal N to this curve at the point P is also shown.

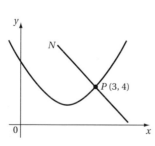

a Find the gradient of the tangent T to the curve at P.

b Hence, show that an equation for N is $2y + x = 11$.

6 Find an equation for the normal to these curves at the given point. Give each answer in the form $ay + bx + c = 0$ for integers a, b and c.

a $y = 9 - x - x^2$ at the point $P(2, 3)$.

b $y = (x^2 - 3)(2x + 1)$ at the point $P(-1, 2)$.

c $y = (x^2 + 1)(x^2 - 8)$ at the point where $x = 2$.

7 **(PS)** The diagram shows the curve $y = 3x^2 - 5x + 7$. The curve crosses the y-axis at point A. The tangent T to this curve at point P is also shown. The x-coordinate of P is 2 and the tangent crosses the y-axis at point Q.

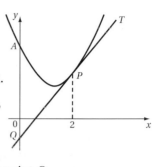

Find the area of triangle PAQ.

8 **(PS)** The diagram shows the curve $y = 3 + 4x - 4x^2$ which passes through the point $P(1, 3)$. The tangent T and the normal N to this curve at point P are also shown. Points A and B are where the tangent T and normal N intersect the y-axis respectively.

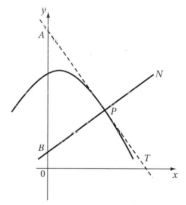

a Show that $AB = \dfrac{17}{4}$.

b Hence, or otherwise,

  i find the area of triangle APB

  ii find the value of $AP \times BP$.

9 **(PS)** A curve has equation $y = x^2 + px + q$ where p and q are constants.

a Find $\dfrac{dy}{dx}$.

The line $y = 5x - 14$ is a tangent to this curve at the point where $x = 4$.

b i Show that $p = -3$.

ii Find the value of q.

c Hence, find an equation for the tangent to this curve at the point where $x = \frac{1}{3}$.

Give your answer in the form $ay + bx = c$ where a, b and c are integers.

10 The diagram shows the curve $y = x^2 - 5x + 10$. Also shown is the normal N to the curve at the point P. The x-coordinate of P is 3. N intersects the curve again at point Q.

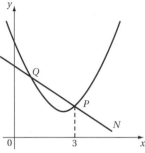

a Show that an equation for N is $y = 7 - x$.

b Use algebra to find the coordinates of point Q.

11 a Find the coordinates of the stationary points on these curves.

i $y = 2x^2 - 12x + 13$

ii $y = 5 - 4x - 4x^2$

b Hence sketch, on separate diagrams, each curve in part **a**. On each sketch, label the y-intercept and the stationary point with their coordinates.

12 The diagram shows the curve $y = 2x^3 + 3x^2 - 12x + 5$. The curve has stationary points at P and Q.

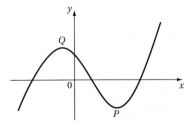

a Show that the x-coordinate of any stationary point on this curve satisfies the equation $x^2 + x - 2 = 0$.

b Hence, find the coordinates of P and the coordinates of Q.

13 a Find the coordinates of any stationary points on the curves with these equations.

i $y = x^3 - 12x + 10$

ii $y = x^3 - 6x^2 + 9x - 2$

iii $y = 3x^2 + 7 - x^3$

iv $y = x^2(2x - 9) + 3$

v $y = (x + 1)^2(2 - x) + 1$

vi $y = 4x^3 - 9x^2 + 6x - 4$

b Hence sketch, on separate diagrams, the curves with equations in part **a**. You do not need to find the coordinates of the points where each curve crosses the x-axis.

> **Handy hint**
>
> Use the y-intercept of each curve to help you complete the sketch.

14 Prove that the curves with these equations do
(PS) not have any stationary points.

a $y = x^3 + x + 1$

b $y = x^3 + 3x^2 + 6x - 1$

c $y = x^3 + kx^2 + k^2x + 1$ where k is a non-zero constant.

15 Tom and Shazia are studying A Level Maths.
(PS) Tom correctly states that every equation $ax^3 + bc^2 + cx + d = 0$, where a, b, c and d are real numbers and $a \neq 0$, has at least one real solution.

Shazia says, 'Therefore, the curve C with equation $y = x^4 + x^3 + x^2 + x + 1$ has at least one stationary point.'

a Explain why Shazia is correct.

Tom also states that the curve C crosses the x-axis at least once.

b By multiplying each side of the equation $x^4 + x^3 + x^2 + x + 1 = 0$ by $(x - 1)$ and simplifying, show that Tom is incorrect.

9 Practice: Integration

9.1 The rules of integration

1 Find these integrals.

  a $\int 6x^2 + 4x + 1 \, dx$ **b** $\int 2x^4 - 3x^3 + 5x^2 \, dx$

  c $\int 7 - 3x - \frac{1}{2}x^2 \, dx$ **d** $\int \frac{3}{2}x^5 + \frac{2}{3}x^3 \, dx$

2 By expanding the brackets and simplifying each integrand, find these integrals.

  a $\int (3x + 2)(x + 2) \, dx$

  b $\int (2x + 1)(x^2 - 2x + 1) \, dx$

  c $\int (4x - 3)^2 \, dx$

  d $\int x^2(3x + 2)^2 \, dx$

3 Express $\int \left(\frac{1}{2}x + \frac{2}{3} \right)^2 dx$ in the form $\frac{1}{36}x(px^2 + qx + r) + c$ where p, q and r are integers to be stated and c is an arbitrary constant.

4 Evaluate these definite integrals.

  a $\int_0^2 4x + 3 \, dx$ **b** $\int_1^2 6x^2 + 4 \, dx$

  c $\int_2^4 x^3 - 2x \, dx$ **d** $\int_{-2}^1 4x - 3x^2 \, dx$

  e $\int_{-3}^{-1} \frac{1}{2}x^3 - 5x + 1 \, dx$ **f** $\int_1^2 3x^5 - 2x^3 - 2x \, dx$

5 By expanding the brackets and simplifying each integrand, evaluate these definite integrals.

  a $\int_1^4 x(x - 2) \, dx$ **b** $\int_1^3 (3x + 1)(2x - 1) \, dx$

  c $\int_0^{\frac{1}{2}} (3x - 1)^2 \, dx$ **d** $\int_{-2}^0 x\left(\frac{1}{2}x + 1 \right)^2 dx$

  e $\int_{\frac{1}{3}}^{\frac{1}{2}} 3x(1 - 3x) \, dx$ **f** $\int_{-3}^2 (2x^2 + 1)(x + 3) \, dx$

6 It is given that $\int_2^4 Ax + 2 \, dx = 34$, where A is a constant.

  a Find the value of A.

  b For this value of A, evaluate $\int_0^3 (Ax + 2)^2 \, dx$.

7 Show that $\int_{-a}^a 2x^3(3x^2 - 2) \, dx = 0$ for all real numbers $a > 0$.

8 By simplifying each integrand, find these integrals.

  a $\int \frac{2x^2 + 3x}{x} \, dx$ **b** $\int \frac{x^2 + 5x + 6}{x + 2} \, dx$

  c $\int \frac{3x(x^2 - 4)}{x - 2} \, dx$ **d** $\int \frac{x^2(2x^2 - 7x - 4)}{2x + 1} \, dx$

9 It is given that $\int_1^k 2x - 3 \, dx = 6$, where k is a positive constant.

  a Show that $k^2 - 3k - 4 = 0$.

  b Find the value of k and hence evaluate
$$\int_{-k}^k 2x - 3 \, dx.$$

10 Find the value of the positive constant k for which

Handy hint
For part **d**, form and solve a quadratic in k^2.

a $\int_1^k 3x^2 - 4 \, dx = 3$

b $\int_3^k \frac{1}{3}x^2 - 2x \, dx = 6$

c $\int_k^{k+1} 2x + 5 \, dx = 14$

d $\int_0^k 4x(x+2)(x-2) \, dx = 9$

11 Find the exact value of these definite integrals.

a $\int_0^{\sqrt{2}} x(3x - 1) \, dx$ **b** $\int_1^{\sqrt{3}} x^2(4x + 3) \, dx$

c $\int_{\sqrt{3}}^3 (x - 1)(x - 3) \, dx$

Handy hint
In part **d**, $x^4 - 10x^2 + 9$ is a quadratic in x^2.

d $\int_{\sqrt{2}}^{\sqrt{3}} \frac{4(x^4 - 10x^2 + 9)}{x + 3} \, dx$

(PS) 12 It is given that $\dfrac{dy}{dx} = (5x^2 - 2)(x^2 + 1)$, where y is a function of x.

Handy hint
Integration is the reverse of differentiation.

a Find an expression for y in terms of x and an arbitrary constant.

Given further that when $x = 1$, $y = 2$

b evaluate $\int_0^2 y \, dx$.

(PS) 13 It is given that $\int f(x) \, dx = 3x^4 - 5x^2 + c$, where c is an arbitrary constant.

Find $\int \frac{1}{x} f(x) \, dx$.

Handy hint
Differentiation is the reverse of integration.

9.2 Applying integration

1 a Use integration to find the area of the shaded region **R** in each of these diagrams.

i

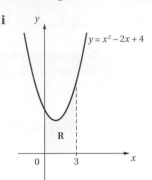

$y = x^2 - 2x + 4$

ii

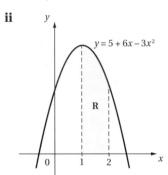

$y = 5 + 6x - 3x^2$

iii

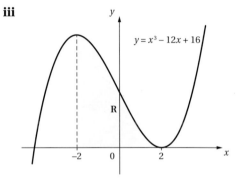

$y = x^3 - 12x + 16$

iv

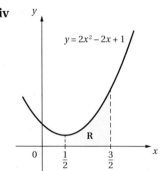

$y = 2x^2 - 2x + 1$

b Check your answer to part **a i** by calculating an estimate for the area of **R** using 3 trapeziums of equal width.

Handy hint

For **1b** use the curve equation to find the side lengths of each trapezium.

c Check your answer to part **a iii** by using 4 trapeziums of equal width. Comment on your findings.

2 Find the area of the region bounded by these curves and the x-axis between the given values of x. In each part, the entire region lies above the x-axis.

a The curve with equation $y = x^3 - 3x^2 + 3x - 1$, between $x = 1$ and $x = 2$.

b The curve with equation $y = x^4 - 2x^2 + 1$, between $x = -1$ and $x = 1$.

c The curve with equation $y = 2x(5 - x)$, between $x = 1$ and $x = 3$.

3 The diagram shows the curve with equation $y = (x - 1)(3 - x)$.

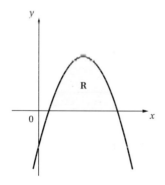

a Write down the values of x where this curve crosses the x-axis.

b Find the area of the shaded region **R** bounded by this curve and the x-axis.

4 The diagram shows the curve with equation $y = 2 + x - x^2$.

R is the shaded region bounded by this curve and the x-axis.

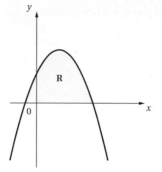

a Find the area of **R**.

b Show that the y-axis divides **R** into two regions with areas in the ratio $7 : 20$.

5 The diagram shows the curve with equation $y = x^2 - 5x + 7$ and the line $y = x - 1$.

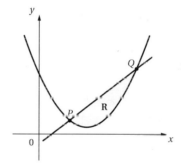

This line and curve intersect at points P and Q.

a Find the coordinates of P and Q.

R is the region bounded between this curve and line.

b Find the area of **R**.

6 Find the area of the region **R** in these diagrams.

a

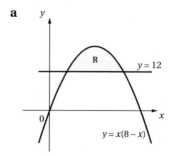

b

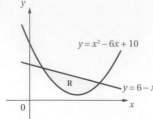

$y = x^2 - 6x + 10$

R

$y = 6 - x$

c Find the area of the shaded region bounded by the curve with equation $y = (x + 1)(4 - x)$ and the x-axis, and which excludes region **R**.

9 The diagram shows the curve with equation
PS $y = x^3 + 8$.

R is the shaded region bounded by this curve, the x-axis and the y-axis.

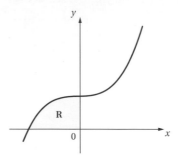

R

PS **7** The diagram shows the curve with equation $y = 5 + 4x - x^2$, and the line $2y - x = 6$. P is an intersection point of this curve and line. The x-coordinate of P is positive and this curve passes through point Q on the x-axis.

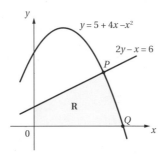

$y = 5 + 4x - x^2$

$2y - x = 6$

P

R

Q

a Find the coordinates of point P.

b Find the area of the shaded region **R** bounded by this curve and line and the coordinate axes, as shown on the diagram.

PS **8** The diagram shows the curve with equation $y = x^2 - 3x + 4$ and the curve with equation $y = (x + 1)(4 - x)$.

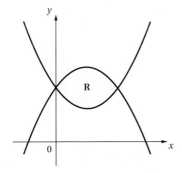

R

a Find the coordinates of the points where these two curves intersect.

   R is the shaded region bounded by these two curves.

b Find the area of **R**.

a Find the area of **R**.

b Sketch the graph of $y = (x - 2)^3 + 8$.

c Using your answer to part **a**, or otherwise, find the value of $\int_0^2 (x - 2)^3 \, dx$.

10 The diagram shows the curve with equation
PS $y = 4 + 6x - 9x^2$ which passes through the point P with x-coordinate $\frac{2}{3}$.

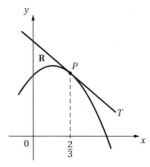

R

P

T

$\frac{2}{3}$

Also shown is the tangent T to this curve at the point P.

R is the shaded region bounded by this curve, T and the y-axis.

Find the area of **R**.

1 Answers: Surds and indices

1.1 Surds

1 **a** $7\sqrt{5}$ **b** $6\sqrt{6}$ **c** $2\sqrt{2}$
  d 20 **e** $-16\sqrt{2}$ **f** $2\sqrt{7}$

2 **a** $5\sqrt{5}$ **b** $\sqrt{2}$ **c** $-4\sqrt{3}$

3 3

4 **a** $4+3\sqrt{2}$ **b** $1-3\sqrt{3}$ **c** $1+5\sqrt{2}$
  d $10+4\sqrt{6}$ **e** $2+7\sqrt{3}$ **f** $5+2\sqrt{6}$

5 **a** $2-\sqrt{3}$ **b** $6+2\sqrt{3}$ **c** $6-2\sqrt{3}$

6 **a** $11+4\sqrt{7}$ **b** $5+2\sqrt{3}$ **c** $-5-3\sqrt{3}$

7 2

8 **a** **i** $2\sqrt{2}$ **ii** $2\sqrt{10}$

  b Because $D = -3b^2 < 0$ for any $b \neq 0$ and the square root of a negative number does not have a real value.

9 **a** $5\sqrt{3}$ **b** $6\sqrt{5}$ **c** $8\sqrt{3}$ **d** $9\sqrt{6}$

10 **a** 2

  b $8+2\sqrt{14}, a=8, b=2$

11 **a** $2+2\sqrt{3}$

  b Hint: Show the three numbers obey Pythagoras' theorem $c^2 = a^2 + b^2$, where $c = 2 + 2\sqrt{3}$.

  c $3+2\sqrt{3}, a=3, b=2$

12 **a** $6+2\sqrt{5}$

  b Hint: Use Pythagoras' theorem.

  c Hint: Use the result of part **a** to show $QR = 2 + 2\sqrt{5}$. Answer: Perimeter and area both equal $10 + 6\sqrt{5}$.

1.2 Indices

1 **a** 2^7 **b** 2^9 **c** 2^{10} **d** 2^{24}

2 **a** 2^{-4} **b** 4^{-3} **c** 4^2 **d** $9^{\frac{5}{2}}$ **e** $8^{-\frac{2}{3}}$ **f** 16^{-2}

3 $\frac{5}{4}$

4 **a** $\frac{4}{3}$ **b** $\frac{3}{2}$ **c** $\frac{2}{3}$ **d** 7

5 **a** $2x^{-1}$ **b** $\frac{1}{2}x^{-3}$ **c** $3x^{\frac{3}{2}}$
  d $\frac{1}{4}x^{\frac{2}{3}}$ **e** $2x^{-\frac{1}{2}}$ **f** $3x^{\frac{2}{3}}$
  g $\frac{1}{2}x^{-\frac{1}{2}}$ **h** $10x^{\frac{1}{4}}$

6 **a** True (add indices)

  b False (for example, $n = 2, a = 3, b = 2$)

  c False (for example, $a = 2, m = 2, n = 1$)

  d True (add indices, $a^0 = 1$)

  e False (for example, $a = 2, n = 3$)

7 **a** $3x + 2x^{-2}$ **b** $1 - \frac{3}{2}x^{-1} + \frac{1}{2}x^{-2}$

8 **a** $2x - 1 - x^{-1}$ **b** $9x^{-1} + 12x^{-2} + 4x^{-3}$
  c $x^{\frac{3}{2}} + 3x^{\frac{1}{2}} - 6x^{-\frac{1}{2}}$ **d** $4x^{-2} + 4x^{-\frac{3}{2}} + x^{-1}$

9 **a** $6 + 13x^{-1} + 6x^{-2}$

  b $13x^{-1} = \dfrac{13}{x}$ and $6x^{-2} = \dfrac{6}{x^2}$ so as x increases, these fractions approach, but never equal, 0 So $6 + 13x^{-1} + 6x^{-2}$ approaches, but never equals, 6.

  c Yes: the point $(-\frac{6}{13}, 6)$, found by solving $\dfrac{(3x+2)(2x+3)}{x^2} = 6$.

2 Answers: Algebra 1

2.1 Basic algebra

1 **a** $5a + 3$ **b** $-5b + 18$ **c** $10a$ **d** $2a^2 + 3b^2$

2 **a** $xy(2x + y)$ **b** $2x^2y^2(5x - 2y)$

   c $3x^3yz(xy + 2z)$ **d** $3xy(4x^3y + 2xy - 3)$

3 **a** $4a^4b^2$ **b** $9a^3b^2(3b^4 + a)$

   c $4a^2b^4(2a + b)(2a - b)$

4 **a** $Q = \dfrac{P}{3} - 4 \ \left(\text{or } \dfrac{P - 12}{3}\right)$

   b $B = \dfrac{2A + 1}{3}$ **c** $T = \dfrac{R + 3}{2}$

   d $D = \dfrac{2C - 5}{12}$ **e** $V = 9U^2 - 2$

   f $N = \sqrt[3]{\dfrac{2M}{\pi}} + 1$

5 **a** $r = \sqrt[3]{\dfrac{3V}{4\pi}}$ **b** 3 cm

6 **a** Hint: The quarter-circle has area $\frac{1}{4}\pi(2x)^2$.

   b $x = \sqrt{\dfrac{A}{4 - \pi}}$

   c Hint: The quarter-circle has perimeter $\frac{1}{4}(2\pi)(2x)$.

7 **a** Hint: Start by expressing t in terms of m using $\tan \hat{A} = \dfrac{\text{opposite}}{\text{adjacent}}$.

   b $m = \sqrt{\dfrac{n^2 - 2}{2}}$

8 **a** $x = \pm\sqrt{y} - 3$

   b $x = 1 \pm \sqrt{\dfrac{y + 1}{4}} \ \left(\text{or } x = 1 \pm \frac{1}{2}\sqrt{y + 1}\right)$

   c $x = \dfrac{5 \pm \sqrt{3y}}{2}$

9 **a** Hint: Use the rule $\dfrac{a + b}{c} = \dfrac{a}{c} + \dfrac{b}{c}$.

   b $Q = \dfrac{3}{P - 2}$

10 **a** $B = \dfrac{2}{1 - A}$ **b** $D = \pm\sqrt{\dfrac{4}{C - 1}}$

   c $F = \sqrt[3]{\dfrac{5}{E + 4}}$

11 **a** $B = \dfrac{2A}{A - 1}$ **b** $D = \dfrac{2 - 3C}{2C - 1}$ **c** $F = \pm\sqrt{\dfrac{3 - E}{E - 1}}$

12 **a** $x + 3$ **b** $2x^2 + 4$ **c** $3x$ **d** $-x^2$

13 **a** $x^2 + \frac{4}{3}x + 2, A = 1, B = \frac{4}{3}, C = 2$

   b $2x - \frac{3}{2}x^{-1}, A = 2, B = \frac{3}{2}$

   c $2x^2 + \frac{1}{3}x^{-2} + 3, A = 2, B = \frac{1}{3}, C = 3$

   d $4x^3, A = 4, n = 3$

2.2 Solving linear equations

1 **a** 5   **b** 8   **c** 3   **d** -3

2 **a** 5   **b** $\frac{1}{5}$ **c** -4

3 **a** 4   **b** 8   **c** 6

4 **a** 3   **b** $\frac{1}{2}$ **c** $\frac{1}{4}$

5 **a** Hint: Substitue $x = 6$ and $y = 3$ into the equation.

   b 8

6 **a** $x = 2, y = 3$ **b** $x = 7, y = -1$

   c $x = \frac{1}{2}, y = -\frac{3}{2}$

7 **a** $2(5x - 1)$ **b** Hint: First find the value of x.
Answer: Area = 68 cm².

8 **a** Hint: The circumference of a semi-circle radius x is πx.

   b 18 800 cm² (3 sig. figs)

9 **a** 7   **b** 12   **c** 3

2.3 Forming expressions

1 a $6x + 8$

  b Hint: Length $BE = \frac{1}{2}(2x + 4) = x + 2$.

    Area of trapezium $= \frac{1}{2}(a + b)h$, where a, b are parallel sides and h is the height.

2 a i $10(x + 1)$ **ii** $5x(x + 3)$ **b** $350\ \text{cm}^2$

3 a $2\pi(x + 3)$

  b Hint: The area of larger circle is $\pi(x + 3)^2$.

4 a Hint: The quarter-circle has circumference $\frac{1}{2}\pi r$.

  b $\frac{1}{4}r^2(\pi - 2)$

5 a $P = 4x$ **b** $x = \sqrt{2}y,\ k = \sqrt{2}$

6 a $2x + y = 24$

  b Hint: Start by rearranging $2x + y = 24$ to make y the subject.

  c $64\ \text{m}^2$

7 a $V = 6x^3$

  b Hint: The cuboid has 6 faces. The base has area $2x^2$, and so on.

  c $\frac{3}{11}$ **d** $88\ \text{cm}^2$

8 a $48 - 4x^2$

  b Hint: Label the sides of the tray with its dimensions. For example: The height of the tray is x cm.

  c $22.5\ \text{cm}^3$

9 a $S = 2xy + 8x + 8y$

  b Hint: Use Volume = base × width × height.

  c $S = 8\left(x + \frac{4}{x} + 1\right)$

10 a $V = \pi r^2 h$

  b Hint: Express r in terms of h. Then use Pythagoras' theorem where L is the hypotenuse.

11 a Hint: Use Diagrams 1 and 2 to express r in terms of L.

  b Hint: Use Diagram 1 to find the curved surface area of the closed cylinder. Then use the formula for the area of a circle radius r.

3 Answers: Coordinate geometry 1

3.1 Straight-line graphs

1 a

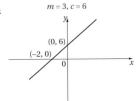

$m = 3, c = 6$

b

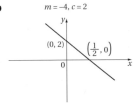

$m = -4, c = 2$

c

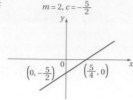

$m = 2, c = -\frac{5}{2}$

$\left(0, -\frac{5}{2}\right)$ $\left(\frac{5}{4}, 0\right)$

d

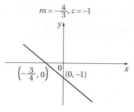

$m = -\frac{4}{3}, c = -1$

$\left(-\frac{3}{4}, 0\right)$ $(0, -1)$

2 a $m = 2, c = -1$ **b** $m = \frac{3}{2}, c = 1$

   c $m = \frac{4}{3}, c = -\frac{1}{3}$ **d** $m = -2, c = 12$

3 a common gradient 2

   b common gradient $\frac{3}{2}$

   c common gradient $-\frac{1}{2}$

4 a $3 \times -\frac{1}{3} = -1$

   b $-\frac{3}{2} \times \frac{2}{3} = -1$

   c $-\frac{5}{3} \times \frac{3}{5} = -1$

5 a

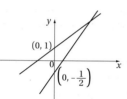

$(0, 1)$

$\left(0, -\frac{1}{2}\right)$

   b $\frac{3}{2}$

6 a

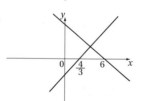

$\frac{4}{3}$ 6

   b $\frac{14}{3}$

7 a $2y + x + 3 = 0$ **b** $3y + 2x - 1 = 0$

   c $4y + 3x - 2 = 0$ **d** $6y - 4x + 15 = 0$

8 A = (1), B = (4), C = (2), D = (3)

9 a

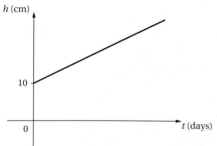

h (cm)

10

t (days)

---

   b h – intercept 10 means the sunflower was 10 cm tall when it was planted.

    Gradient 3.5 means the sunflower grew 3.5 cm per day.

   c 20 weeks

   d For example: according to the model, h increases without limit over time. It is also unlikely that the sunflower grows at a constant rate.

10 a

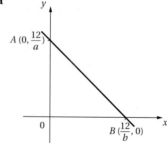

v (m/s)

7

0 5 t (seconds)

   b i v-intercept 7 means the athlete crossed the finish line with speed 7 m/s.

     t-intercept 5 means it took the athlete 5 seconds after crossing the finish line to stop.

    ii Gradient -1.4 means the athlete was slowing down (decelerating) at a rate of 1.4 m/s per second after crossing the finish line.

   c Hint: Area under a speed–time graph = distance travelled.

    Answer: Total distance run = 117.5 metres.

11 a

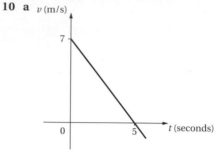

y

$A\left(0, \frac{12}{a}\right)$

0 $B\left(\frac{12}{b}, 0\right)$ x

   b Hint: Use area of triangle = $\frac{1}{2} \times$ base $\times$ height.

   c Hint: Solve the equations $ab = 8, b = 2a$ simultaneously. Answers: $a = 2, b = 4$.

   d Hint: The area of triangle AOB equals $\frac{1}{2} \times AB \times h$, where h is the shortest distance from O to L. You can use Pythagoras to find the distance AB.

3.2 Finding the equation of a line

1 a i $y = 2x - 2$ **ii** $y = 5x - 8$

iii $y = \frac{1}{2}x + \frac{11}{2}$ **iv** $y = -3x - 13$

b $2y - x = 11$

2 a $-\frac{1}{2}$ **b** $2y + x - 6$ **c** $(6,0)$

3 a $y = 4x - 5$ **b** $y = -7$ **c** $2y + 3x = 8$

4 a 13   **b** 5   **c** 2

5 a $y = \frac{3}{2}x + 3$ **b** 3

6 Hint: Find the coordinates of A and B. Then use Pythagoras' theorem on triangle OAB, where O is the origin.

7 a -2

b B does not lie on this line. If you substitute $x = -7$ into the equation, $y = -2(-7) - 3$
$= 14 - 3$
$= 11$
which is not the y-coordinate of B.

8 a Hint: Substitute $x = \frac{5}{2}, y = \frac{1}{2}$ into the equation.

b $\frac{9}{2}$ **c** $\frac{81}{16}$

9 a Hint. Substitute $x = 1, y = k$ into the equation.

b $3y + x = 7$

10 a Hint: Simplify $\frac{4 - 4k}{k - 1}$.

b $y = -4x + 4k + 4$

c Hint: Find the coordinates of R in terms of k. Answer: $k = 2$.

11 a $3y + x = 19$ (or any equivalent form)

b Hint: Find the gradient of CM where M is the midpoint of AB. You will find a sketch helpful.

c Hint: Find the distance CM.

12 a Hint: Start by showing the gradient of the line is $\frac{2}{3}$.   **b**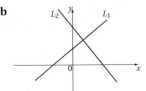

c Hint: Show that the values $x = 6$ and $y = 5$ satisfy both equations.

d 39

3.3 Mid-points and distances

1 a $(6, 4)$ **b** $(2, 3)$ **c** $\left(-\frac{3}{2}, \frac{15}{2}\right)$ **d** $\left(2, \frac{11}{12}\right)$

2 a $\left(\frac{k + 2}{2}, k + 2\right)$ **b** $(-k, k + 3)$ **c** $(2k, 2k + 1)$

3 $(3, 8)$

4 Hint: First find the coordinates of B. The coordinates of D are $(-1, 13)$.

5 a Hint: Solve $\frac{p + 14}{2} = 8, q = 19$

b $(5, 7)$ **c** 15

6 Hint: First find the coordinates of the midpoint of BD. The coordinates of C are $(1, 11)$.

7 a $\sqrt{29}$ **b** $3\sqrt{10}$ **c** $2\sqrt{26}$ **d** $\frac{5}{4}$

8 Hint: Calculate each distance AB, AC and BC. You will see that exactly two of these distances are equal.

9 a $\left(6, \frac{1}{2}\right)$ **b** Hint: Find the midpoint of AB.

10 a Hint: Find the length AC.   **b** $(-3, -15)$

11 a Hint: Use the distance formula to find the length of each side of the triangle.

b It is a right-angled triangle (with the right-angle at vertex B). This is the **converse** of Pythagoras' theorem.

c Hint: The angle in a semi-circle is 90° so AC is a diameter of this circle.
Answer: Area $= \frac{25}{2}\pi$ square units.

12 a Hint: Solve the equation $\sqrt{(p - 1)^2 + 9} = p$.

b 5π

3.4 Intersections of lines

1 a i 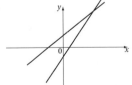   **ii** $(6, 15)$

b i   **ii** $\left(\frac{1}{2}, \frac{7}{4}\right)$

c i

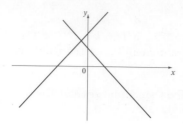

ii $\left(-1, \frac{7}{6}\right)$

2 a $(3, -2)$ **b** $(-2, 0)$ **c** $(-4, -4)$

3 a Hint: Show that trying to solve the equations leads to an impossible result.

b The lines are parallel.

c

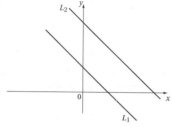

4 a Lines intersect at $(3, -2)$.

b Lines do not intersect (parallel, gradient $\frac{3}{5}$).

c Lines intersect at $(0, \frac{1}{3})$.

5 a i $\left(2, \frac{7}{2}\right)$ **ii**

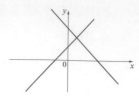

b i $(0, 4)$ **ii**

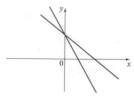

c i $(5, 3)$ **ii**

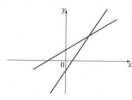

6 a $A: x = 1$, $B: x = \frac{7}{2}$

b $\frac{15}{4}$

7 a $\frac{7}{3}$ **b** 28

8 a $\left(\frac{2}{3}, 1\right)$ **b** $4y + 3x = 6$

9 a Hint: Use $y = mx + c$ where $m = \frac{1}{4}$. **b** $(2, 5)$.

c Hint: First find the equation of the line through AB. Distance $BC = 2\sqrt{13}$.

4 Answers: Algebra 2

4.1 Solving a quadratic equation by factorising

1 a $x = 3, 5$ **b** $x = -7, 2$ **c** $x = 3$

d $x = -\frac{5}{2}, 2$ **e** $x = \frac{1}{2}, 5$ **f** $x = -3, \frac{4}{3}$

2 a $x = -6, 3$ **b** $x = \frac{3}{2}, 2$ **c** $x = -3, 4$

d $x = -\frac{1}{4}, 2$ **e** $x = -2, 8$ **f** $x = -\frac{4}{3}, 4$

3 a $p = -14, 4$ **b** $q = 4, 18$ **c** $r = -\frac{15}{2}, 3$

4 a Hint: Replace every x in the equation with 2.

b $x = 13$

5 $x = \frac{1}{2}$

6 a $y^2 - 5y + 4 = 0$ **b** $y = 1, 4: x = \pm 1, \pm 2$

7 a $x = \pm 2, \pm 3$ **b** $x = 1, 9$

   c $x = 0, 3$ **d** $x = -1, 1$

8 a Hint: Expand and simplify the equation.
$(2x - 5)(x - 3) = 1$

   b $x = \dfrac{7}{2}$ (reject $x = 2$ as this results in the rectangle having negative lengths)

   c 5 km

9 Hint: Use Pythagoras' theorem $BD = \dfrac{13}{2}$ cm.

10 Hint: Simplify and solve the equation.
$3x(x + 1) = 2(4x + 1)$

   Area = 18 cm²

11 a Hint: Use Area = $\dfrac{1}{2}(a + b)h$ where a and b are the lengths of the horizontal sides and h is the height of the trapezium.

   b $\dfrac{161}{8}$ cm²

   c Hint: Use congruent triangles and Pythagoras' theorem to find PQ.

12 a 40 cm   **b** 30 cm

13 a Hint: Use Pythagoras's theorem to find BC^2 and factorise the answer.

   b 336 cm²

4.2 Using the quadratic formula

1 a $x = -3 \pm 2\sqrt{2}$ **b** $x = -2 \pm \sqrt{7}$

   c $x = 3 \pm \sqrt{6}$ **d** $x = \dfrac{1 \pm \sqrt{17}}{4}$

   e $x = \dfrac{2 \pm \sqrt{10}}{3}$ **f** $x = \dfrac{5 + \sqrt{31}}{2}$

2 a $x = 1 \pm \sqrt{7}$ **b** $x = \dfrac{-2 \pm \sqrt{13}}{3}$

   c $x = 5 \pm 2\sqrt{7}$

3 To 2 decimal places:

   a $x = 0.70, 4.30$ **b** $x = 0.45, 2.22$

   c $x = -1.78, 4.78$

4 a Two real roots   **b** No real roots

   c One real root

5 a Hint: Discriminant = $4(k^2 + 1)$.

   b $2k$ **c** -1

6 a $x = \dfrac{k \pm \sqrt{k^2 - 8}}{2}$ **b** $x = \sqrt{k}, \dfrac{1}{2}\sqrt{k}$

   c $x = -\dfrac{1}{k}$ **d** $x = \dfrac{k \pm 3}{2}$

7 Hint: The discriminant $16 - 4(1)(2p)$ must be negative.

8 a $q > \sqrt{56} = 7.48\ldots$ so $q = 8$

   b $x = \dfrac{-4 \pm \sqrt{2}}{2}$

9 a Hint: Solve $\pi r^2 + 20r - 240 = 0$.

     $r = 6.1$ cm (1 decimal place)

   b Hint: Calculate $\dfrac{\pi r^2}{240} \times 100$.

10 Hint: Use discriminants to show that $16 < c < 18$.

     Answer: $p = -6 - \sqrt{2}, q = -6 + \sqrt{2}$

11 a Hint: Simplify $(p - 1)^2 + (p - 3)^2 = 36$.

   b Hint: Solve the equation $p^2 - 4p - 13 = 0$ to find p then use the area of a trapezium formula.

     Answer: Area = $11 + 3\sqrt{17}, m = 11, n = 3$

12 a Hint: Form and solve the equation.
     Answer: $(p - 2)^2 + (p - 4)^2 = 16$

   b Hint: Use $\text{Grad}_{AB} = \text{Grad}_{AC}$.

     Answer: $q = 6 + \sqrt{7}$

   c Hint: Use $AB + BC = AC$ where $AB = 4$

13 a $7 + 4\sqrt{3}, a = 7, b = 4$

   b Hint: Use Pythagoras' theorem.

   c Hint: $x^2 = 7 + 4\sqrt{3}$ (reject $x^2 = 7 - 4\sqrt{3}$ as then $x < 1$).

     Answer: $x = 2 + \sqrt{3}$

4.3 Further equation solving

1 a $x = -1, y = 2 : x = -2, y = 1$

   b $x = -\dfrac{9}{5}, y = -\dfrac{13}{5} : x = 1, y = 3$

   c $x = -\dfrac{1}{11}, y = \dfrac{19}{11} : x = -1, y = -1$

2 a $A\left(\dfrac{2}{5}, \dfrac{11}{5}\right), B(2, -1)$

   b Hint: Simplify $\sqrt{\left(2 - \dfrac{2}{5}\right)^2 + \left(-1 - \dfrac{11}{5}\right)^2}$

3 a $x = -\dfrac{4}{3}, y = -\dfrac{11}{3} : x = 2, y = 3$

   b $x = 2, y = \dfrac{1}{2} : x = -1, y = -1$

   c $x = \dfrac{17}{8}, y = -\dfrac{7}{4} : x = -2, y = 1$

4 a $(0, 10)$ **b** $(-4, 2)$ **c** 20

5 a $(3, 1)$

 b The line is a tangent to the circle at the point $(3, 1)$

6 a Hint: Show that trying to solve the equations leads to a quadratic equation with a negative discriminant.

 b The line does not intersect or touch the circle.

7 a $x = -4, -2, 3$ **b** $x = -3, 0, 2$

 c $x = -2, \frac{2}{3}, 5$ **d** $x = -1, 2, 3$

 e $x = -4, 0, 1$ **f** $x = 0, \frac{3}{4}$

 g $x = -\frac{2}{3}, 0, 4$ **h** $x = 0, \frac{3}{2}$

8 a $x^2 + y^2 = 5, 4x + 2y = 10$. Answers are $x = 2, y = 1$

 b Larger square: Area = 4 m². Smaller square: Area = 1 m².

9 a $x^2 + y^2 = 36, x + y = 8$. Answers $AB = 4 + \sqrt{2}$, $AC = 4 - \sqrt{2}$

 b 7 cm²

10 a Hint: Use Pythagoras' theorem.

 b $L = 2x - 1$

 c Hint: Solve $L^2 = 5x^2 - 8x + 4$ and $L = 2x - 1$ simultaneously.

 Answer: Area = 12 km²

5 Answers: Coordinate geometry 2

5.1 Transformations of graphs

1 a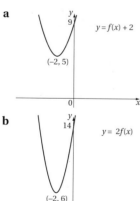

 b

 c

d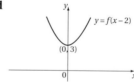

2 a i $(4, 0)$ **ii** $(3, 2)$ **b**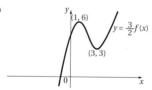

3 a Stretch, scale factor $\frac{1}{a}$ parallel to x-axis from the origin.

 b 4   **c** $\left(\frac{3}{2}, 0\right), \left(-\frac{1}{4}, 0\right)$

 d i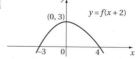

ii

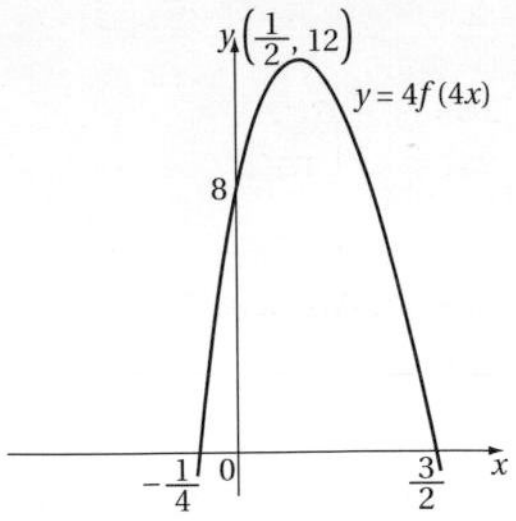

4 a i

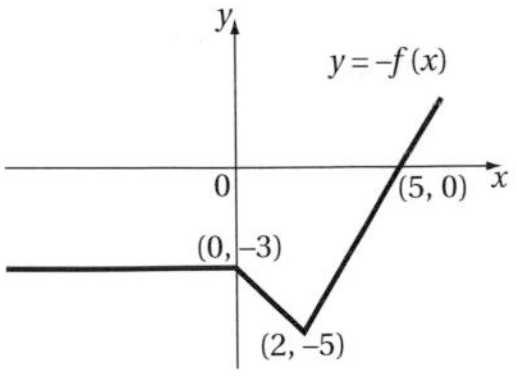

ii

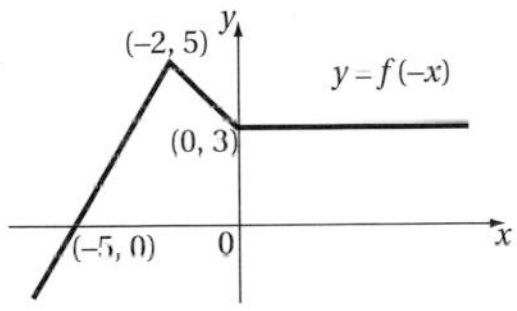

b

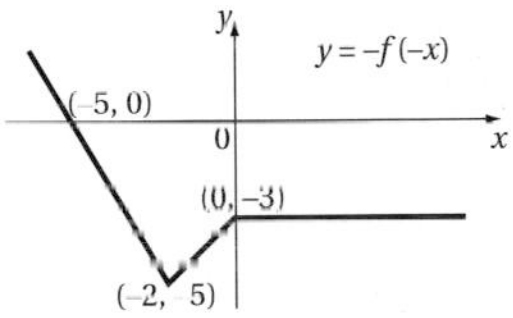

c Rotation 180° about the origin.

5 a i

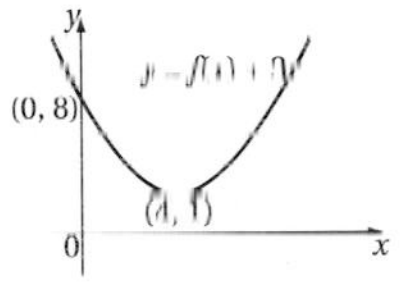

ii

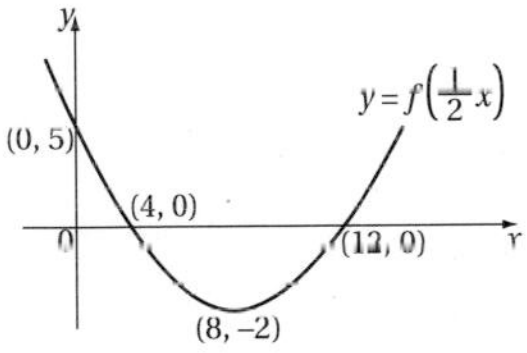

iii

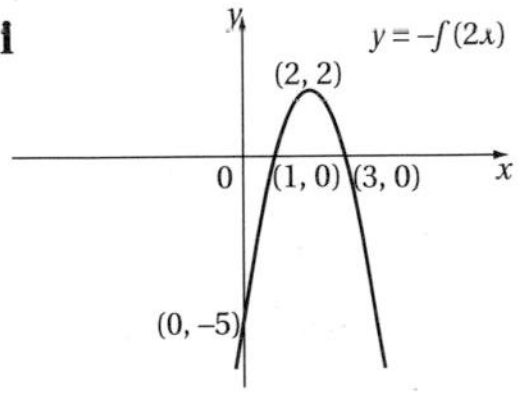

iv

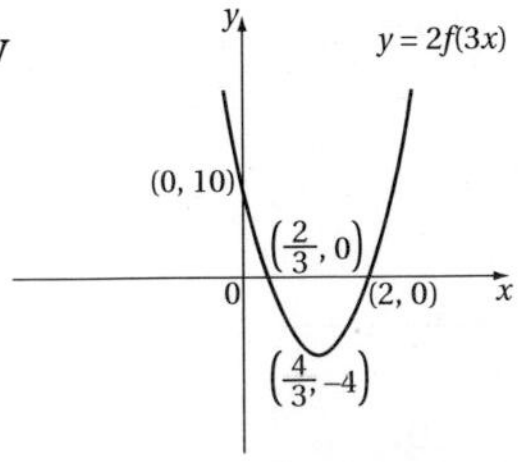

b 2 **c** (2, 0), (0, 2)

5.2 Sketching curves

1 a

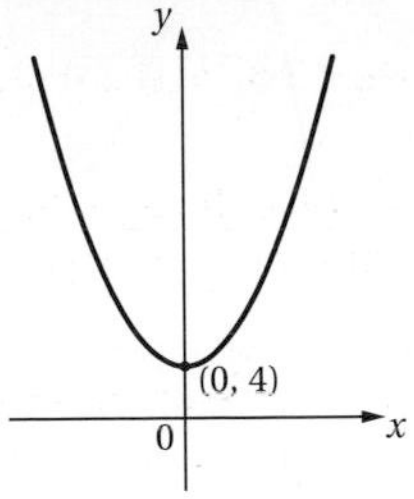

b

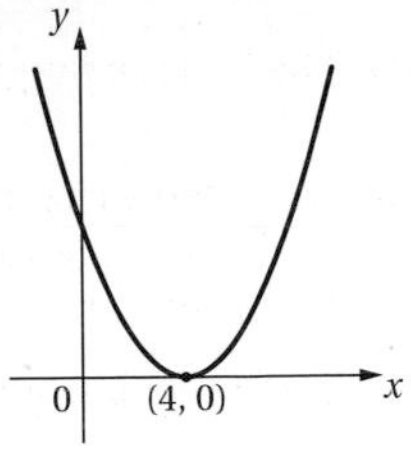

c

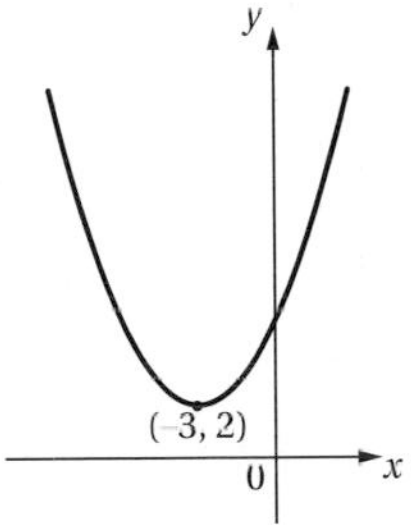

2 a

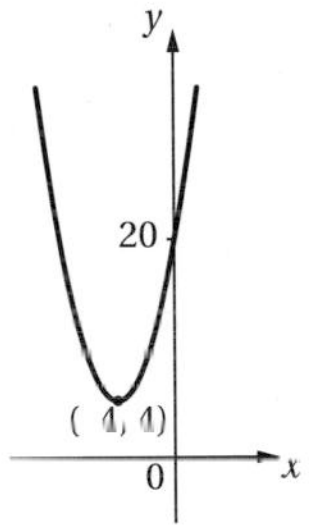

b

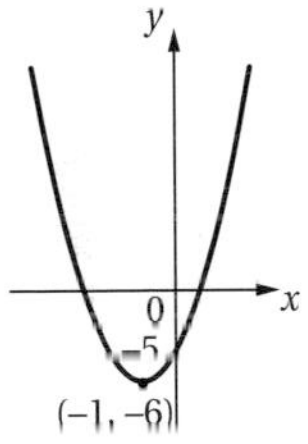

c

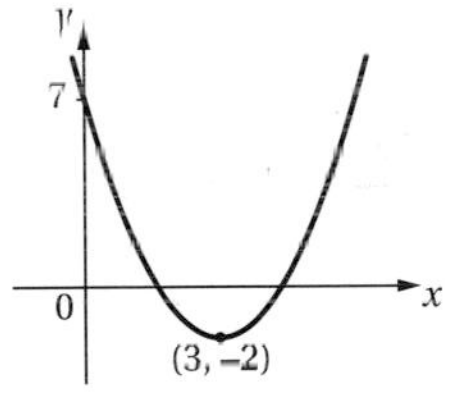

d

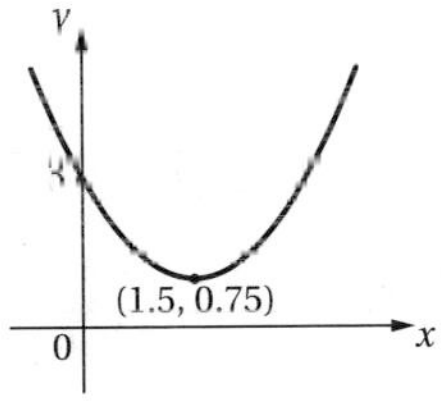

3 a

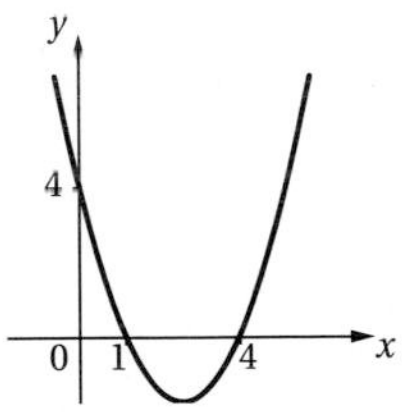

b

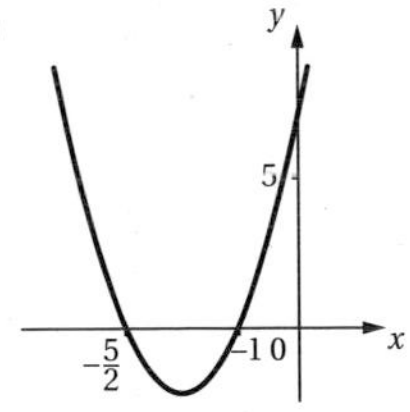

c

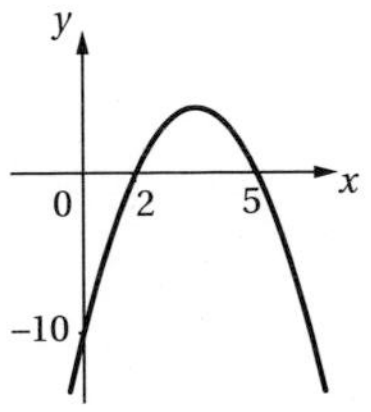

d

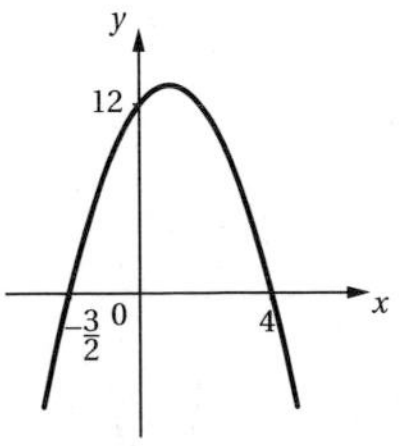

4 a   **b**

c 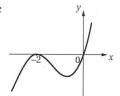 **d**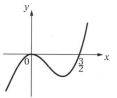

5 a 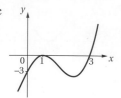 **b**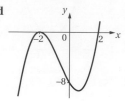

c 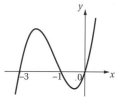 **d**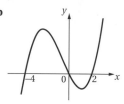

6 a Hint: The curve equation is $y = (x - 3)^2 - 4$.
Expand and simplify this equation.

Answers: $b = -6, c = 5$

b $y = (x - 1)(x - 5)$

7 a $x^3 - 13x + 12$

b Hint: Factorise $x^2 + x - 12$.

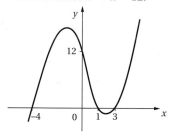

8 a A $\leftrightarrow$ Figure 4 , B $\leftrightarrow$ Figure 3 , C $\leftrightarrow$ Figure 1

b Possible equation for Figure 2: $y = (x + 1)^2(x - 1)$

9 a $b = -4, c = 3, p = 2, q = -1$

b $b = -5, c = 6, q = -\dfrac{1}{4}, r = 3$

c Hint: Using symmetry, $\dfrac{(r + 3) + r}{2} = 3$.

Answers: $b = -6, c = \dfrac{27}{4}, q = -\dfrac{9}{4}, r = \dfrac{3}{2}$

d Hint: Show that $r = 2q - 6$ and that $q^2 - q = 6r$.
Answers: $b = -8, c = 12, q = 4, r = 2$

---

10 a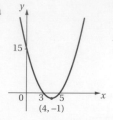

b Hint: Translate the graph in part **a** up the y-axis
so that its turning point is on the x-axis.
Answer: $k = 1$

c Hint: Show that the minimum point has
coordinates $\left(\dfrac{a + b}{2}, -\left(\dfrac{b - a}{2} \right)^2 \right)$.

5.3 Intersection points of graphs

1 a $A(2, 5), B(6, 13)$

b Hint: Simplify $\sqrt{(6 - 2)^2 + (13 - 5)^2}$

2 a $(-2, 6), (4, 0)$ **b** $(0, 7), (-3, 1)$

c $(-1 + \sqrt{2}, 4\sqrt{2}), (-1 - \sqrt{2}, -4\sqrt{2})$

d $(3 + \sqrt{3}, 2 + \sqrt{3}), (3 - \sqrt{3}, 2 - \sqrt{3})$

3 a $A(1, 0), B(6, 0)$ **b** $C(2, 4), D(5, 4)$ **c** 16

4 a $A(0, 5), B(0, -13)$

b Hint: Solve $2x^2 - 4x + 5 = 8x - 13$. Answer: (3, 11)

c Hint: The base is the distance AB and the height
is the x-coordinate of P.

5 a Hint: Show that solving $x^2 - 4x + 7 = 1 - 2x$
leads to a quadratic equation with negative
discriminant.

b The line and curve do not intersect.

c $(x - 2)^2 + 3$ **d**

6 a 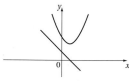 **b** $k < 4$

7 a $(0, 1), (-1, 3), (-2, 5)$

b $(0, 1), \left(\dfrac{3}{2}, -\dfrac{23}{4} \right) (3, 10)$

8 a $A(1, 14)$, $B(6, 9)$

b Hint: Start by finding the equation of L.
$(-0.2, 7.8)$, $(5.2, 13.2)$

9 Hint: Start by finding the coordinates of A and B.
Answer: Area of $R = 36$ square units.

10 a $(-1, 0)$, $(0, 1)$, $(5, 96)$

b Hint: Find the roots of each curve by using the factors of each equation.

The cubic curve has roots -1 and 1 (repeated)

The quadratic curve has roots -1, $-\frac{1}{3}$

The y-intercept of both curves is 1

6 Answers: Trigonometry

6.1 Trigonometry and triangles

Unless stated otherwise, answers are given to 3 significant figures where appropriate.

1 a 4.43 cm   **b** 11.4 cm   **c** 11.6 cm   **d** 2.99 cm

2 a Hint: Use the sine rule $\dfrac{\sin \hat{B}}{b} = \dfrac{\sin \hat{A}}{a}$.   **b** 9.82 cm

3 a Hint: Rearrange $\dfrac{\sin \hat{B}}{2x} = \dfrac{\sin 30}{x}$ to make $\sin \hat{B}$ the subject.

b Hint: Find $\sin^{-1}(1)$.   **c** $\sqrt{3}x$

4 a $\cos \hat{C} = \dfrac{a^2 + b^2 - c^2}{2ab}$

b $\hat{B} = \cos^{-1}\left(\dfrac{a^2 + c^2 - b^2}{2ac}\right)$

5 a 2.97 cm   **b** 12.4 cm

6 a $\hat{C} = 83.2°$ **b** $\hat{A} = 53.8°$, $\hat{B} = 43.0°$

7 a Hint: Use the sine rule on triangle ACD

b Hint: Use the sine rule on triangle BCD

Answer: Angle $DBC = 90°$

c Hint: Use right-angled trigonometry to find AB

Answer: Perimeter $= 6 + 6\sqrt{3}$ cm

8 a Hint: Use the cosine rule to find BC in terms of x

b Hint: The perpendicular from C to AB bisects the base.

Answer: $k = 27$

9 a Hint: Use the cosine rule on triangle BAD.

b 21.2°   **c** 83.8°

10 a Hint: Use the cosine rule on triangle BCD, where $BC = 2x$.

b 11.1 cm

11 a Hint: Use the cosine rule to form a quadratic equation in k.

  b Angle $ACB = 28°$ (nearest degree), angle $BAC = 32°$ (nearest degree)

12 a Hint: Use the sine rule on triangle ADB.

  b $93.0°$

  c Opposite angles in a cyclic quadrilateral sum to $180°$.
  Since $\hat{A} + \hat{C} = 134.1° + 93.0° = 227.1$, $ABCD$ cannot be a cyclic quadrilateral.
  Hence, all four points cannot lie on a common circle.

  d Hint: Start by finding angle CDB and then use the cosine rule on triangle ADC.

6.2 The area of any triangle

Where appropriate, answers are given to 3 significant figures unless stated otherwise.

1 a 29.7 cm^2 **b** 27.2 cm^2 **c** 17.2 cm^2

2 23.3 cm^2

3 a Hint: Rearrange the sine rule $\dfrac{\sin \hat{R}}{10} = \dfrac{\sin 150°}{30}$ to make $\sin \hat{R}$ the subject.

  b $9.59°$ **c** Hint: Use the formula $\frac{1}{2} qr \sin \hat{P}$.

4 10.8 cm^2

5 a Hint: Rearrange $c^2 = a^2 + b^2 - 2ab \cos \hat{C}$ to make $\cos \hat{C}$ the subject.

  b 26.8 cm^2

6 a 22.6 cm^2

  b Hint: The sector CAB has area $\left(\dfrac{45°}{360°}\right) \times \pi \times (8)^2$.

7 a 9.00 cm^2 **b** 3.01 cm

  c Hint: Start by using the cosine rule to find the length of the side PR.

8 a Hint: Find an expression for the area of triangle PQR in terms of x. Answer: $PQ = 6 \text{ cm}$

  b Hint: Use the cosine rule $p^2 = q^2 + r^2 - 2qr \cos \hat{P}$.

  c Hint: Use Area $= \frac{1}{2} \times$ base $\times$ perpendicular height. Answer $= 4.84 \text{ cm}$

9 Hint: Find the area of the sector and the area of the triangle.

10 a Hint: Use the sine rule to find the length of side AB. Answer: Area of plate $= 43.8 \text{cm}^2$

  b 35.6 cm

11 a Hint: Start by finding the length AE by using Pythagoras' theorem on triangle ABE. Then use the cosine rule on triangle AEF.

  b $\dfrac{5}{2} \text{ cm}^2$

  c Hint: Start by calculating angle BEA using right-angled trigonometry. Then find angle CEF.

  d $\dfrac{104}{25} \text{ cm}^2$

6.3 Solving a trigonometric equation

Where appropriate, answers are given to 1 decimal place.

1 a $x = 0°, x = 180°, x = 360°$ **b** $36.9°$ **c** $143.1°$

2 a $180°$ **b** $x = 134.4°, x = 225.6°$

3 a $90°$ **b** $x = 14.5°, x = 165.5°$

  c $x = 30°, x = 150°, x = 390°, x = 510°$

4 a $x = 0°, x = 360°$ **b** $x = 48.2°, x = 311.8°$

  c $x = 135°, x = 225°, x = 495°, x = 585°$

5 a Hint: Both answers are equal to $\dfrac{1}{\sqrt{2}}$.

  b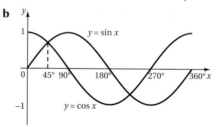

  c $45°$ (see sketch)   **d** $225°$

6 a Hint: Rearrange the sine rule $\dfrac{\sin \hat{B}}{b} = \dfrac{\sin \hat{A}}{a}$ to make $\sin \hat{B}$ the subject.

  b $34.8°$ **c** $145.2°$ **d** e.g.

7 a $x = 30°, x = 150°$ **b** $x = 70.5°, x = 289.5°$

  c $x = 45°, x = 135°$ **d** $x = 75.5°, x = 284.5°$

8 a Hint: Apply a suitable translation to the graph of $y = \sin x$.

b $x = 78.5°$, $x = 281.5°$

c $x = 48.6°$, $x = 131.4°$

9 a $-30°$ **b** $210°$

c $x = 233.1°$, $x = 306.9°$

10 a Hint: Sketch the graph of $y = \sin x$ and the line $y = \dfrac{5}{4}$ on the same diagram.

b i Not possible as $\cos x \leqslant 1$ for all x.

ii $270°$

iii Not possible as $\sin x \geqslant -1$ for all x.

iv $x = 41.4°$, $x = 318.6°$

11 a Stretch, scale factor $\dfrac{1}{a}$, along the x-axis from the origin.

b $x = 23.6°$, $x = 156.4°$

c i $x = 11.8°$, $x = 78.2°$ **ii** $x = 5.9°$, $x = 39.1°$

iii $x = 15.7°$, $x = 104.3°$

12 a $x = 112.0°$, $x = 248.0°$

b i $x = 37.3°$, $x = 82.7°$ **ii** $x = 56.0°$, $x = 124.0°$

13 a $x = -286.3°$, $x = -73.7°$, $x = 73.7°$, $x = 286.3°$

b $x = -319.5°$, $x = -220.5°$, $x = 40.5°$, $x = 139.5°$

c $x = -135.6°$, $x = -44.4°$, $x = 224.4°$, $x = 315.6°$

d $x = -246.4°$, $x = -113.6°$, $x = 113.6°$, $x = 246.4°$

7 Vectors

7.1 The magnitude and direction of a vector

1 a $3\sqrt{2}$, $\theta = 45°$ **b** 13, $\theta = 112.6°$

c $2\sqrt{10}$, $\theta = 341.6°$ **d** $2\sqrt{2}$, $\theta = 210°$

2 a $8\sqrt{2}$, $\theta = 45°$ **b** $2\sqrt{2}$, $\theta = 135°$

c 16, $\theta = 180°$ **d** $\dfrac{16}{15}$, $\theta = 90°$

3 a $\mathbf{p} = \begin{pmatrix} 6\cos 30° \\ 6\sin 30° \end{pmatrix} = \begin{pmatrix} 3\sqrt{3} \\ 3 \end{pmatrix}$ **b** $\mathbf{q} = \begin{pmatrix} -2 \\ 2\sqrt{3} \end{pmatrix}$

c $\mathbf{r} = \begin{pmatrix} 0 \\ -5 \end{pmatrix}$ **d** $\mathbf{s} = \begin{pmatrix} 1 \\ -1 \end{pmatrix}$

4 a Hint: work out direction of $\mathbf{q}$ minus direction of $\mathbf{p}$.

b $2\sqrt{5}$

5 Answers to part **a** rounded 1 decimal place.

a i 9.1, $\theta = 6.3°$ **ii** 17.0, $\theta = 310.2°$

iii 18.4, $\theta = 112.4°$

b $\mathbf{r} = -6\mathbf{i} - 2\mathbf{j}$

6 a $a = 12$ **b** $b = 2$ **c** $30°$

7 a $\mathbf{d} = \begin{pmatrix} \sqrt{6}\cos 45 \\ \sqrt{6}\sin 45 \end{pmatrix} = \begin{pmatrix} \sqrt{3} \\ \sqrt{3} \end{pmatrix}$ or $\sqrt{3}\mathbf{i} + \sqrt{3}\mathbf{j}$

b $60°$

c $2(1 + \sqrt{3})$

8 a i $4\sqrt{2}\mathbf{i} + 4\sqrt{2}\mathbf{j}$ **ii** $-\dfrac{3\sqrt{2}}{2}\mathbf{i} - \dfrac{3\sqrt{2}}{2}\mathbf{j}$

iii $-2\sqrt{2}\mathbf{i} + 2\sqrt{2}\mathbf{j}$

b Hint: Use Pythagoras to find the lengths UV and VW.

Answer: Perimeter $= 16 + 4\sqrt{5}$

7.2 Position vectors

1 a i $\begin{pmatrix} 3 \\ 2 \end{pmatrix}$ **ii** $\begin{pmatrix} 4 \\ -3 \end{pmatrix}$ **iii** $\begin{pmatrix} \sqrt{3} - 3 \\ \sqrt{3} + 3 \end{pmatrix}$

b $\overrightarrow{OB}$ (magnitude $= 5$)

c

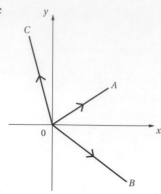

2 a i $\begin{pmatrix} -10 \\ 2 \end{pmatrix}$ **ii** $2\sqrt{26}$

  b i $\begin{pmatrix} 8 \\ 4 \end{pmatrix}$ **ii** $4\sqrt{5}$

  c i $\begin{pmatrix} \sqrt{2} \\ 4 \end{pmatrix}$ **ii** $3\sqrt{2}$

3 a i Hint: Show that $\overrightarrow{BC} = 2 \times \overrightarrow{AB}$.

    ii Hint: show that an equation for the line through A and B is $3y = x + 11$.

  b You probably found the vector approach more straightforward. It has the advantage in telling you that the ratio $AB:BC = 1:2$

4 a $k = -2$ or $k = 6$

  b

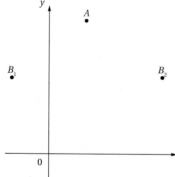

  c Point A

5 a $\sqrt{2}, \theta = 135°$

  b $5\sqrt{2}, \theta = 45°$

  c Hint: first show that $\overrightarrow{OA}$ and $\overrightarrow{OC}$ are perpendicular.

6 a Hint: Use $\overrightarrow{OC} = \overrightarrow{OA} + \overrightarrow{AC}$ and $\overrightarrow{OC} = \overrightarrow{OB} + \overrightarrow{BC}$.

  b $AB = \sqrt{52}$, $BC = \sqrt{32}$, $AC = \sqrt{68}$

  c Hint: Use the cosine rule $b^2 = a^2 + c^2 - 2ac \cos \hat{B}$.

  d Hint: Use $\frac{1}{2}ac \sin \hat{B}$. Area = 20 square units

7 a Hint: Use the definition of a parallelogram.

  b Hint: Use the results $\overrightarrow{AB} = \overrightarrow{DC}$ and $\overrightarrow{BE} = \frac{1}{2}\overrightarrow{BD}$.

  c Hint: Use the result $\overrightarrow{AB} = \overrightarrow{DC}$.

    Answer: E is the midpoint of AC so the diagonals of a parallelogram bisect each other.

8 a Hint: Use $\overrightarrow{AC} = \overrightarrow{OC} - \overrightarrow{OA}$ to find $\overrightarrow{AC}$.

  b $(6, 6)$

  c Hint: Use the cosine rule to find the exact values of $\cos \alpha$ and $\cos \beta$.

9 a Hint: Use the fact that $\overrightarrow{AB}$ is parallel to $\overrightarrow{OC} = \begin{pmatrix} 6 \\ 2 \end{pmatrix}$.

  b Hint: Show $|\overrightarrow{AC}| = 5$ and then solve the equation $\left| \begin{pmatrix} 6k + 1 \\ 2k + 2 \end{pmatrix} \right| = 5$.

    Answer: $B(4, 3)$

  c Hint: Find $|\overrightarrow{AB}|$, $|\overrightarrow{BC}|$ and then use the cosine rule on triangle ABC.

    Answer: Angle $ABC = 135°$

8 Answers: Differentiation

8.1 Estimating the gradient of a curve

1 a 5   **b** 3   **c** 4

2 a i 6 underestimate   **ii** 10 overestimate

    iii 9.66 (3 sig. figs) overestimate

   b Answer **iii**: $\sqrt{2} \approx 1.41$ so of the points Q, R and S, point S is the closest one to point P

3 a 0

   b $(x - 2)^2 + 3$ so coordinates of R are (2, 3)

   c Hint: $\text{Grad}_{PQ} = 3$, $\text{Grad}_{RP} = 1$ so the integer m is such that $1 < m < 3$.

   d 2

4 a 8   **b** −12   **c** −7   **d** 1

5 For example: choose $A(2.9, 7.1025)$ and $B(3.1, 7.4025)$

   Let m = gradient of curve at $x = 3$

   $\text{Grad}_{AP} = 1.475$, $\text{Grad}_{PB} = 1.525$

   $1.475 < m < 1.525$ so $m = 1.5$ (1 decimal place)

6 For example: choose $A(1.99, 0.49246...)$ and $B(2.01, 0.50746...)$

   Let m = gradient of curve at $x = 2$

   $\text{Grad}_{AP} = 0.746...$, $\text{Grad}_{PB} = 0.753...$

   $0.746... < m < 0.753...$ so $m = 0.75$ (2 decimal places)

7 a From the diagram, $m < \text{Grad}_{PQ} = 5$

   b $P(1, 2)$, $Q(1.5, 4.5)$

   c For the lines to intersect at a point with a positive x-coordinate, the tangent must be steeper than the line OQ.

    So $m > \text{Grad}_{OQ}$ where $\text{Grad}_{OQ} = \dfrac{4.5}{1.5} = 3$

   d $m = 4$, $y = 4x - 2$

8.2 The rules of differentiation

1 a Hint: Expand $3(x + h)^2$, then simplify

    $\dfrac{3(x + h)^2 - 3x^2}{(x + h) - x}$.

   b Hint: Expand $(x + h)^2 - 5(x + h)$, then simplify

    $\dfrac{\left((x + h)^2 - 5(x + h)\right) - (x^2 - 5x)}{(x + h) - x}$.

2 a $x^4 + 4x^3h + 6x^2h^2 + 4xh^3 + h^4$

   b Hint: Simplify $\dfrac{(x + h)^4 - x^4}{(x + h) - x}$.

3 a $2(x + 2)$ **b** $8(x + 1)$

   c $3x(x - 4)$ **d** $4x^0(x + 3)$

   e $(3x + 5)(x - 3)$ **f** $(5x^2 + 1)(x^2 + 1)$

4 a 7   **b** −8   **c** 18   **d** $\dfrac{5}{2}$

5 Roots are $\dfrac{1}{2}$ and 3

   At $x = \dfrac{1}{2}$, gradient of curve = 7

   At $x = -3$, gradient of curve = −7

6 a −2

   b Hint: Solve the equation $6x - 2 = 10$.

   c $Q\left(-\dfrac{1}{2}, \dfrac{3}{4}\right)$

7 a Hint: To find the x-coordinate of P, solve the equation $x^2 - \dfrac{4}{3} = \dfrac{8}{3}$.

    $P(2, 0)$, $Q(5, 35)$

   b Hint: Solve the equation $x^2 - \dfrac{4}{3} = \dfrac{35}{3}$

    $R\left(\sqrt{3}, 3\sqrt{13}\right)$

8 Hint: Show that the gradient of the curve at P is 9, then solve the equation $3x^2 + 6x = 9$:
Answer: $Q(-3, 2)$

9 a Hint: Find $\dfrac{dy}{dx}$ and solve the equation $3k + \dfrac{2}{3} = \dfrac{17}{3}$.

  b $\dfrac{64}{3}$

  c From the diagram, the gradient of the curve at $B < \dfrac{64}{3}(= 21.33...)$

  David's answer $\dfrac{43}{2} = 21.5 > 21.33...$ and so he must have made a mistake.

  d Hint: Solve the equation $5x^2 + \dfrac{2}{3}x = 8$ and select the answer which lies between 1 and 2.

  $B\left(\dfrac{6}{5}, \dfrac{109}{25}\right)$

10 a $3x^2$ **b** $6x^2 - 3$ **c** $4x - \dfrac{9}{2}$

  d 1 **e** $\dfrac{2}{3}$ **f** $2x$

11 a Hint: The common denominator is $(x + h)x$.

  b Hint: Simplify $\dfrac{\left(\dfrac{1}{x+h} - \dfrac{1}{x}\right)}{h}$.

  c $\dfrac{1}{x} = x^{-1} \xrightarrow{\text{differentiates to}} (-1)(x^{-2}) = -\dfrac{1}{x^2}$

12 a h

  b Hint: Use the result of part **a** to write $\dfrac{\sqrt{x + h} - \sqrt{x}}{h}$ as $\dfrac{1}{\sqrt{x + h} + \sqrt{x}}$.

  c $\sqrt{x} = x^{\frac{1}{2}} \xrightarrow{\text{differentiates to}} \dfrac{1}{2}\left(x^{-\frac{1}{2}}\right) = \dfrac{1}{2x^{\frac{1}{2}}} = \dfrac{1}{2\sqrt{x}}$

8.3 Applying differentiation

1 a Hint: Find an expression for $\dfrac{dy}{dx}$ and substitute in the value $x = 2$.

  b $y = 3x - 2$

2 a $y = -4x + 2$ **b** $y = 9x - 17$

  c $y = 2x - 1$ **d** $y = -5x + 8$

3 Hint: Solve $(x^2 + 2)(x - 3) = 0$ to find where this curve crosses the x-axis. Tangent equation is $y = 11x - 33$

4 a i $y = 6x - 8$

  ii Hint: Solve $4x^2 - 6x + 1 = 11$ to find a.

  Tangent equation is $y = 14x - 24$

  b $(2, 4)$

5 a 2

  b Hint: Use $Grad_N \times Grad_T = -1$.

6 a $5y - x - 13 = 0$ **b** $2y - x - 5 = 0$

  c $4y + x + 78 = 0$

7 Hint: Start by finding the equation of T. Answer: for area of triangle $PAQ = 12$ square units.

8 a Hint: Find an equation for T and for N.

  b i $\dfrac{17}{8}$ **ii** Hint: Use Area of triangle $= \dfrac{1}{2} \times$ base $\times$ perpendicular height.
  Answer: $AP \times BP = \dfrac{17}{4}$

9 a $2x + p$

  b i Hint: The tangent has gradient 5, so $\dfrac{dy}{dx} = 5$ when $x = 4$.

  ii Hint: The tangent and curve intersect at the point where $x = 4$ $q = 2$.

  c $9y + 21x = 17$

10 a Hint: Start by finding the gradient of the curve at point P.

  b $(1, 6)$

11 a i $(3, -5)$ **ii** $\left(-\dfrac{1}{2}, 6\right)$

  b i   **ii**

12 a Hint: Show that $\dfrac{dy}{dx} = 0$ leads to the equation $x^2 + x - 2 = 0$.

  b $P(1, -2)$, $Q(-2, 25)$

13 a i $(2, -6)$, $(-2, 26)$ **ii** $(3, -2)$, $(1, 2)$

  iii $(0, 7)$, $(2, 11)$ **iv** $(0, 3)$, $(3, -24)$

  v $(1, 5)$, $(-1, 1)$ **vi** $\left(\dfrac{1}{2}, -\dfrac{11}{4}\right)$, $(1, -3)$

  b i   **ii**

  iii **iv**

  v **vi**

14 a Hint: Find $\dfrac{dy}{dx}$ and then show the equation $\dfrac{dy}{dx} = 0$ has no real solutions.

b Hint: Find $\dfrac{dy}{dx}$ and then use the quadratic formula on the equation $\dfrac{dy}{dx} = 0$.

c Hint: Find $\dfrac{dy}{dx}$ and then use the quadratic formula on the equation $\dfrac{dy}{dx} = 0$, leaving the answers in terms of k.

15 a Stationary points on the curve with equation $y = x^4 + x^3 + x^2 + x + 1$ correspond to the real solutions to the equation $\dfrac{dy}{dx} = 0$

Since $\dfrac{dy}{dx} = 4x^3 + 3x^2 + 2x + 1$, the equation $4x^3 + 3x^2 + 2x + 1 = 0$ must have at least one real solution (by Tom's statement).

So the curve C must have at least one stationary point.

b Any real solution to the equation $x^4 + x^3 + x^2 + x + 1 = 0$ must also be a real solution to the equation $(x - 1)(x^4 + x^3 + x^2 + x + 1) = 0$

Expand the brackets and simplify: $x^5 - 1 = 0$

so: $x^5 = 1$

The only real solution to the equation $x^5 - 1$ is $x = \sqrt[5]{1} = 1$

But $x - 1$ is clearly **not** a solution to the equation $x^4 + x^3 + x^2 + x + 1 = 0$

So the equation $x^4 + x^3 + x^2 + x + 1 = 0$ does not have any real solutions. This means the curve C cannot cross the x-axis.

9 Answers: Integration

9.1 The rules of integration

Where it appears, c is an arbitrary constant

1 a $2x^3 + 2x^2 + x + c$

b $\dfrac{2}{5}x^5 - \dfrac{3}{4}x^4 + \dfrac{5}{3}x^3 + c$

c $7x - \dfrac{3}{2}x^2 - \dfrac{1}{6}x^3 + c$

d $\dfrac{1}{4}x^6 + \dfrac{1}{6}x^4 + c$

2 a $x^3 + 4x^2 + 4x + c$

b $\dfrac{1}{2}x^4 - x^3 + x + c$

c $\dfrac{16}{3}x^3 - 12x^2 + 9x + c$

d $\dfrac{9}{5}x^5 + 3x^4 + \dfrac{4}{3}x^3 + c$

3 $\dfrac{1}{36}x(3x^2 + 12x + 16) + c$, $p = 3, q = 12, r = 16$

4 a 14   **b** 18   **c** 48   **d** -15

   e 12   **f** 21

5 a 6   **b** 46   **c** $\dfrac{1}{8}$ **d** $-\dfrac{1}{3}$

   e $-\dfrac{1}{18}$ **f** 6

6 a Hint: $\displaystyle\int Ax + 2\,dx - \dfrac{1}{2}Ax^2 + 2x + c$.

     Answer: $A - 5$

   b 327

7 Hint: You will find the result $(-a)^n = a^n$ for even values of n useful.

8 a $x^2 + 3x + c$ **b** $\dfrac{1}{2}x^2 + 3x + c$

   c $x^3 + 3x^2 + c$ **d** $\dfrac{1}{4}x^4 - \dfrac{4}{3}x^3 + c$

9 a Hint: Simplify $\left[x^2 - 3x\right]_1^k$

   b $k = 4$, $\displaystyle\int_{-4}^{4} 2x - 3\,dx = -24$

10 a 2   **b** 9   **c** 4   **d** 3

11 a $2\sqrt{2} - 1$ **b** $7 + 3\sqrt{3}$

   c $6 - 4\sqrt{3}$ **d** $3 - 4\sqrt{2}$

12 a Hint: Integrate $(5x^2 - 2)(x^2 + 1)$ with respect to x.

   b Hint: Use the given values of x and y to find the value of the arbitrary constant c.

13 Hint: Differentiate $3x^4 - 5x^2 + c$ with respect to x.
Answer: $4x^3 - 10x + c$

9.2 Applying integration

1 a i 12   **ii** 7   **iii** 64   **iv** $\dfrac{7}{6}$

   b Area $\approx \dfrac{1}{2}(4+3)\times 1 + \dfrac{1}{2}(3+4)\times 1 + \dfrac{1}{2}(4+7)\times 1$

           $= 12.5$

     This is an overestimate for the exact area of **R** = (12) because all the trapeziums lie above the curve.

   c Area $\approx \dfrac{1}{2}(32+27)\times 1 + \dfrac{1}{2}(27+16)\times 1$

           $+ \dfrac{1}{2}(16+5)\times 1 + \dfrac{1}{2}(5+0)\times 1$

           $= 64$

     Unusually, the trapezium method has calculated the exact area of **R**.

     (This is because the graph of $y = x^3 - 12x + 16$ has 180° rotational symmetry about its y-intercept.)

2 a $\dfrac{1}{4}$ **b** $\dfrac{16}{15}$ **c** $\dfrac{68}{3}$

3 a $x = 1, x = 3$ **b** $\dfrac{4}{3}$

4 a Hint: Start by finding the roots of the graph.

     Area of **R** = $\dfrac{9}{2}$

   b Hint: Find the value of $\displaystyle\int_{-1}^{0} 2 + x - x^2 \, dx$ and $\displaystyle\int_{0}^{2} 2 + x - x^2 \, dx$

5 a Hint: See Section 5.3.

     Answers: $P(2, 1), Q(4, 3)$

   b $\dfrac{4}{3}$

6 a $\dfrac{32}{3}$ **b** $\dfrac{9}{2}$

7 a $P(4, 5)$

   b Hint: Split **R** into two regions, one of which is a trapezium.

     Area of **R** = $\dfrac{56}{3}$

8 a $(0, 4), (3, 4)$

   b Hint: Find the area bounded by each curve and the x-axis between $x = 0$ and $x = 3$.

     Area of **R** = 9

   c $\dfrac{71}{6}$

9 a 12

   b

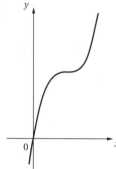

     (The graph of $y = (x - 2)^3 + 8$ is a translation of $y = x^3 + 8$ by the vector $\begin{pmatrix} 2 \\ 0 \end{pmatrix}$)

   c $\displaystyle\int_{0}^{2}(x - 2)^3 + 8 \, dx$ = Area of **R** because a region and its translation have equal areas

      $\displaystyle\int_{0}^{2}(x - 2)^3 \, dx + \int_{0}^{2} 8 \, dx = 12$, where

      $\displaystyle\int_{0}^{2} 8 \, dx = \left[8x \right]_{0}^{2} = 16$

     So $\displaystyle\int_{0}^{2}(x - 2)^3 \, dx = 12 - 16$

                  $= -4$

     'otherwise' : Expand $(x - 2)^3$ and integrate each term.

10 Hint: Start by finding the equation of the tangent T.

     Area of **R** $= \dfrac{8}{9}$

Exam paper

Total = 50 marks

Time allowed: 1 hour

1 a Given that $27^n = \frac{1}{9}$, find the value of n.   [2]

   b Express $\dfrac{5 - 2\sqrt{3}}{4 + \sqrt{3}}$ in the form $a + b\sqrt{3}$ for integers a and b to be stated.

      You must show your working.   [3]

2 Find the value of the constant c for which the equation $4x^2 - 12x + c = 0$ has exactly
   one real root.   [3]

3 The height h (in km) above the ground of an aeroplane t minutes
   after it begins its descent to land is modelled by the equation $h = 2.7 - 0.12t$.

   The diagram shows part of the graph of h against t.

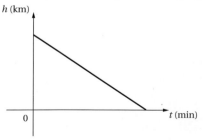

   [*continued*]

   EXAM PAPER

a Describe, in context, what the negative constant in the equation $h = 2.7 - 0.12t$ represents. **[1]**

b **i** Find the coordinates of the point where this line crosses the t-axis. **[2]**

ii Give an interpretation of these coordinates. **[1]**

---

4 The line L_1 has equation $y = 11 - 2x$. The line L_2 has equation $4y - 5x + 8 = 0$.

a Find the coordinates of the point where these lines intersect. **[3]**

b On the diagram provided, sketch L_1 and L_2, labelling each line with its equation.

[2]

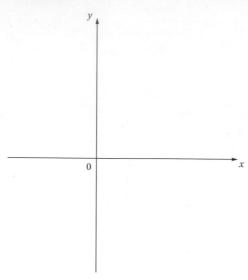

c Find the area of the finite region bounded by these lines and the x-axis.

[3]

5 The diagram shows the curve with equation $y = f(x)$. This curve crosses the x and y-axes at points $A(4, 0)$ and $B(0, 4)$, respectively. Point $C(2, 6)$ is a maximum point on this curve.

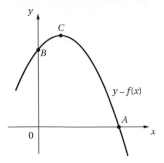

On the diagram provided, sketch the curve with equation $y = f\left(\frac{1}{2}x\right)$.

Label the points A', B' and C' which are the images under this transformation of A, B and C, respectively, with their coordinates.

[3]

6 The diagram shows a triangular metal plate *ABC* welded to a
sector *ACD* of a circle centre *A*, radius 8 cm.

$AB = 3$ cm, $BC = 7$ cm, and angle $BAC = \theta°$.

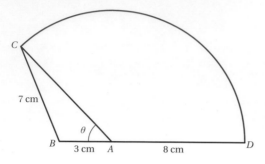

a Find the value of θ. [3]

b Calculate the area of this shape. Give your answer to
1 decimal place. [4]

7 Relative to the origin *O*, the points $A(4, 1)$, $B(0, 6)$, $C(p, 10)$
and *D*, where $p > 4$ is a constant, are the vertices of a square,
as shown in the diagram.

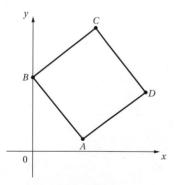

[*continued*]

a Show that $\overrightarrow{AB} = -4\mathbf{i} + 5\mathbf{j}$. [1]

b Find $\left|\overrightarrow{BC}\right|$ in terms of p. [2]

c Show that $p = 5$. [3]

d Find, in $\mathbf{i} - \mathbf{j}$ form, the position vector of D. [3]

8 The diagram shows the graph with equation $y = x^2 - 8x + 19$. Also shown is the tangent T to this curve at the point P with x-coordinate 5 and the line L which intersects this curve at points $A(2, 7)$ and B.

The line L is parallel to T.

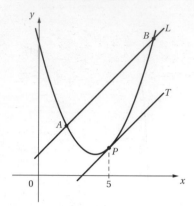

a Find an equation for T. [4]

b Show that an equation for line L is $y = 2x + 3$. [2]

c Find the area of the region bounded by this curve, the line T and the line L. [5]

EXAM PAPER